T0257848

Spectroscopy: Advanced Concepts

Spectroscopy: Advanced Concepts

Edited by **Jason Penn**

New York

Published by NY Research Press,
23 West, 55th Street, Suite 816,
New York, NY 10019, USA
www.nyresearchpress.com

Spectroscopy: Advanced Concepts
Edited by Jason Penn

International Standard Book Number: 978-1-63238-424-9 (Hardback)

Printed in the United States of America.

Contents

Preface

This book aims to highlight the current researches and provides a platform to further the scope of innovations in this area. This book is a product of the combined efforts of many researchers and scientists, after going through thorough studies and analysis from different parts of the world. The objective of this book is to provide the readers with the latest information of the field.

The onset of a new era of science has been marked by developments in spectroscopy and its functions. Spectroscopy scientists from various domains had been encouraged to share and discuss information that has been included in this book. It talks about latest developments in technology, functions and other advancements of spectroscopy. This book will augment the understanding of researchers about complications of different spectroscopic approaches and urge professionals and beginners to contribute part of their future research in understanding relevant mechanisms and applications of chemistry, physics and material sciences. Studies on analytical methods for determination of polyphenols, spectrophotometry and chemiluminometry, quality control of herbal medicines, flow-injection spectrophotometric analysis of Iron (II), Iron (III), and spectrophotometry as a tool for dosage sugars in nectar crops etc. are covered extensively in this book.

I would like to express my sincere thanks to the authors for their dedicated efforts in the completion of this book. I acknowledge the efforts of the publisher for providing constant support. Lastly, I would like to thank my family for their support in all academic endeavors.

Editor

Other Spectroscopy

Spectrophotometry as a Tool for Dosage Sugars in Nectar of Crops Pollinated by Honeybees

Vagner de Alencar Arnaut de Toledo[1,*],
Maria Claudia Colla Ruvolo-Takasusuki[2], Arildo José Braz de Oliveira[3],
Emerson Dechechi Chambó[1] and Sheila Mara Sanches Lopes[3]
[1]*Animal Science Department,*
[2]*Cell Biology and Genetics Department,*
[3]*Pharmacy Department, Universidade Estadual de Maringá, Maringá, Paraná*
Universidade Estadual de Maringá, Maringá,
Brazil

1. Introduction

The pollination by honeybees is important to the best performance of several crops. In this interaction plant-insect there is a change of reward between both organisms, and the sugar concentration in the nectar is a keyword. The spectrophotometry allows analyzing the type and the quantities of sugar in the nectar of flowers, and identifying varieties that are more attractive for pollinators.

The nectar is the reward for several pollinators, and the principal is the honeybee *Apis mellifera*. The nectar is produced from sap of phloem by active secretion that results in a solution of sugars like sucrose, fructose and glucose in varied proportion depending on the vegetal.

Besides sugars, other compounds of the nectar has importance for the coevolution between plants and their pollinators like amino acids, proteins, lipids and alkaloids and these may be toxic for visitors and, however, these compounds may have a role of protection against animals that withdraw nectar of flowers without an efficient pollination.

Several researches are carried out to evaluate the effect of crop pollination using honeybees and consequently the increase of productivity in agriculture. The visit and hoarding of nectar and pollen allows rise in grain production, or tasteful fruit with symmetric format.

The study of sugar from floral nectary is important for identification if a rise or decrease in quantity or nectar quality. The plant may secret a little bit of nectar, but with high sugar concentration, or unlike, secret more quantities, but with low sugar concentration. These differences in nectar may vary depend on pollinator visitation. However, the frequency of honeybees that visit flowers may contribute for rising nectar production like change the sugar proportion.

---

* Corresponding Author

Other factors that must be considered are secreted sugars: sucrose, glucose and fructose. The quantities of them may vary depending on variety and type of vegetal. Honeybees have preference for nectar with more sucrose concentration. Sugars present in nectar are related with honey quality that will be produced by honeybees and, finally will be commercialized.

The association of beekeeping and agriculture provide a rise in profits as for farmer as for beekeeper. However, the quantity and/or quality of the sugars in floral nectar like the pollination of cultivated crops by honeybees have an economic and social role well-established and significant currently.

The proposal of this chapter is perform a review about spectrophotometry in sugar dosages of nectar of the main cultivated crops and show the importance of this tool (spectrophotometry) to improve crop production by honeybee pollination contributing as for agriculture as for beekeeping.

1.1 Pollination and plant reproduction

The pollination process occurs in spermatophitic plants consists in pollen grain transfer to the stigma, which is the receptive part of feminine flowers of superior vegetal. The pollen grains are structures that contain the reproductive male cells, they are produced and storaged in anthers (part of male organ of flowers) until the deiscence, moment wherein are released. The indispensable factor in pollination process and that denotes the success is the need of ovule fertilization and subsequent fruit set and seeds formation.

The pollination of flowers beside the diaspore dispersion is a fundamental process in reproductive success of vegetal species. The involved animals in this process have an important role (Buckmann & Nabhan, 1997), then the efficiency in the pollination process means rise in disponibility of food to the human being and animals.

The structure of plant (monoic or dioic) varies to the size and anatomic and physiological characterists of flower and their position in the plant, may occur the autopollination or cross pollination. This cross pollination provides a rise in gene flow between plants, spreading them and the results are favorable (Malerbo-Souza et al., 2004).

The animals that carried out the pollen transfer of anthers to the stigma flowers are known pollinators, and can be insects like bees, bettles, flies, butterflies, wasps and moth; birds - hummingbirds and parakeets; and small mammals - bats, rodents, and marsupials (Malagodi-Braga, 2005). Among pollinators, animals of Insecta class are the most important, and in the order Hymenoptera you can find the major number of them. Honeybees are the most important pollinators available in the nature.

Unlike of other insects that visit flowers only to collect its own food, honeybees visit a bigger quantity of flowers, besides of food to own survival, they harvesting pollen and nectar to feed their larvae and storaging (Müller et al., 2006). Futhermore, the higher efficiency of honeybees as pollinators is as much as by their number in the nature, as by better adaptation to floral complex structures like mouthparts and adapted body to imbibe the nectar of flowers and harvesting pollen, respectively (Proctor et al., 1996). The bees of Apidae family have higher distinction because their morphological traits are representative, like special structure to load pollen - corbicula, located in tibia of hind leg, similar to basket

– pollen is loaded in this structure in association with nectar or oil, absence of ventral scopa and long tongue (Teixeira & Zampieron, 2008).

Each group among animals that visit flowers is associated to some particular type of floral reward, that is the morphological traits of flowers reflect adaptation to diversified pollinators. The contemporaneous interactions between plants with flowers and insects can be because of long and closer coevolutionary relation (Backer & Hurd, 1968; Prince, 1997). This process of coevolution or the interaction between plants and pollinators is based on a system of mutual dependence. This system was detailed by the first time by Christian Konrad Spengel (1750-1816), in which plants show their rewards like nectar, pollen, oil and resins by floral arrangement, colour, size and odour of flowers, while the pollinators in change of provided resources by plants transfer pollen between flowers increasing the gene flow and promoting the diversification of the species, named by this as key mutualism (Morgado et al., 2002).

Plants, year by year, specialized in attract more efficient pollinators and make transportation of their reproductive cells, and therewith could be benefited with cross pollination. The disposition of flowers is an important factor that can be isolated in the branches or grouped in the same floral axis forming inflorescence, colour, odour, size, nectar, oil, pollen and resins.

The floral rewards provided by Angiosperms are required to attract pollinators, nectar seems the most searched in crop cultivated commercially, however, in searching by this reward many animals, mainly worker honeybees *Apis mellifera* have pollen adhered to their body, and so, later deposit accidentally their loads on the stigma of other flower of the same specie, performing indirectly the cross pollination.

Assays carried out in Marechal Cândido Rondon, Paraná, Brazil, with sunflower make clear worker honeybees that hoarding nectar are more frequent (mean 2.28 honeybees/capitulum) than pollen foragers (0.40 honeybees by capitulum) in anthesis period and schedule of higher visitation in the crop (Chambó et al., 2011). Other experimental results make clear more frequency of nectar foragers than pollen foragers in sunflower crop (Paiva et al., 2002; Teixeira & Zampieron, 2008).

Moreover, it must be considered that quantity of honeybee visitors to different species of superior vegetal can be related to concentration and volume of nectar in the flowers during all day (Pham-Delegue et al., 1990). Experiments with attractiveness sunflower genotypes show significative difference in relation to the number of honeybee visitation, mean of 3.40 (genotypes Helio 360 and Aguará) and 1.60 (genotype Multissol) visits of *A. mellifera* by capitulum in third day of anthesis (Chambó et al., 2011, in press). The researchers did not tested the concentration and volume secreted during all day, but assign to these causes the difference of sunflower genotypes assayed. In hybrid of ornamental coloured sunflower BRS-OASIS had an increase ($p < 0.05$) in four times in number of honeybee visitors using sucrose solution in two concentrations 5% and 7.5% as attractive. These solutions were pulverized on sunflower capitulum in relation to concentration of 2.5% (Martin et al., 2005).

Despite of efficiency of pollination process, it depends on numerous received visits by pollinators (Schirmer, 1985). Vidal et al. (2010) studying the pollination and set fruit in *Cucurbita pepo* by honyebees reported that percentage of set fruit was maximum (100%)

when the flowers received 12 visits of *A. mellifera*, corresponding to load of 1.253 pollen grains deposited on the stigma. In comparision, with two visits of honeybees, 174 pollen grains were deposited on the stigma, and the set fruit only of 50%. Other important factor is the attraction by pollen and nectar of male flower was arised by opening grade of nectary pore of flower (Vidal et al., 2010).

Models of pollen transfer depends on the specific pollinator rate about pollen remotion of anthers and deposition on the stigmas. The pollinators have a high remotion and low deposition (HRLD) of pollen on the stigma of flowers will benefite a plant wherever it is not better available pollinator. In case of pollinators have high remotion of pollen of anthers and high deposition (HRHD) of pollen on the stigmas also visit a plant population, the visits of pollinators HRLD can reduce the total pollen transfer. The HRLDs parasite the plants, displace the pollen grains that would delivery by HRHDs. When two visitors remove equal quantities of pollen, the pollinator more efficient will be that with higher delivery rate. In case of different quantities of remotion, the better do not depends on the deposition rates only, but another variables including schedule of visitation to pollen deposition (Thomson & Goodell, 2002).

The volume and sugar concentration of nectar, important factor in attraction of pollinators alos are known by varing between plant species and affect the answer of pollinator (Lanza et al., 1995). Besides, different varieties of the same species can range to the sugar concentration in the nectar (Free, 1993). In some vegetal species, for example, *Curcubita pepo,* the periodical remotion of nectar of the flower by pollinators do not arise the total volume of produced nectar by plant, then the nectar secretion is not stimulated or inhibited by successed harvesting of this reward. However, the nectar remotion of flowers can reduce or stimulate the process of secretion in several plant species.

The pollinator that drag out the spent time in flowers to nectar hoarding, specially if it is available, increase the probability of pollen deposition and, consequently, the pollination can be well successed. Nevertheless, cultivated commercially crops that do not stimulate the nectar production after several visits of pollinators can have an advantage in pollination process, so the number of honeybee visits to flowers is positively correlated to the nectar secretion in all flower duration (Vidal et al., 2006). The evaluation of nectar secretion rate is an important component in ecological studies related to the pollination process, mainly in that about flower-insect interaction.

2. Use of spectrophotometry in nectar analysis of plants

The quantitative analysis using spectrophotometric methods are widely used because have a good sensitivity, low analysis cost, easily handle, accessible equipments, justifying their application and efficiency for utilization in quantification analysis of several compounds.

The main spectrophotometric techniques used in sugar determination are based on reactions of these carbohydrates with colorimetric reagents, forming a colored complex that can be detected and quantified in a spectrophotometer.

The techniques used for total sugar quantification are anthrone method (Trevelyan & Harrison, 1952 in: Yemm & Wills, 1954), phenol-sulfuric (Dubois et al., 1956) and also can be used the 3-5-dinitrosalicylic acid - DNS modified (Miller, 1959). For reducing sugar determination, also can be used the technique of DNS and the reaction with p-hydroxybenzoic acid hydrazide - *PAHBAH* (Blakeney & Mutton, 1980).

For specific determination of some sugars also are available some other methods. Fructose can be determinated by cysteine/tryptophan/sulphuric acid method (Messineo & Musarra, 1972) and resorcinol method (Roe et al., 1949). Sucrose can be determined by anthrone method after destruction of monosaccharides with KOH (Sala Jr. et al., 2007). Besides these methods, the glucose, fructose and sucrose can be selectively determined by methods using enzymatic reactions (Moernan et al., 2004; Amaral et al., 2007).

3. Total sugar determination

3.1 Phenol-sulphuric method (Dubois et al., 1956)

This method is based on the fact that simple or complex sugar and their derivatives, including methyl esters with free reducing groups or potentially free, when treated with phenol and concentrated sulphuric acid will generate a yellow-orange colour, the reaction is sensible and this colour is stable. The method is simple, quick, sensible, and the results are reproductible.

The changing in colour of solution is measured in visible range and proportional to the quantity of sugar inside the sample.

The carbohydrates are hydrolyzed under heating in strongly acid pH, this reaction produces furan derivative compounds, that when condensed with phenolic compounds produce coloured substances (Figure 1).

The total sugar concentration is determined by spectrophotometry in 490 nm wavelength. Sensitivity of this method range from 10 to 100µg of total sugar and the quantification is made from calibration curve using glucose or pentose as standard and calculation are performed by equation of the linear regression obtained from calibration curve.

3.1.1 Methodology

Reagents: phenol solution 5% (w/v) and concentrate sulphuric acid (95%, p/v).

Get an aliquot of 100µg.mL^{-1} from the sample, add 0.5mL of phenol solution 5%, shake in vortex, add 2.5mL, shake again and keep in water-bath 25°C for 15 minutes. After this period, read the absorbance in spectrophotometer at 490nm.

3.2 Anthrone method (Dreywood, 1946)

The anthrone is the reduction product of anthraquinone and was recognized first as specific reagent for several carbohydrates by Dreywood (1946) because sugar solution in concentrate sulphuric acid form the blue-greenish colour characteristic and since anthrone has been used widely as suitable and specific reagent for sugar colorimetric dosage.

The anthrone reaction is based on hydrolitic and dehydrating action of concentrate sulphuric acid on carbohydrates, in which glicosidic linkages are broken releasing free reducing sugar that are dehydrated and converted to furfural by pentoses and hydroxymethylfurfural by hexoses (Figure 1).

These substances are condensed with anthrone to hydroxyanthracene (9,10-dihydro-9-oxo anthracene) forming a blue-greenish product that have absorption maximum at 620nm.

The sensitivity of this method is from 0 to 100µg.mL^{-1}. Glucose solution is used as standard to build calibration curve and get the straight line equation to quantify samples.

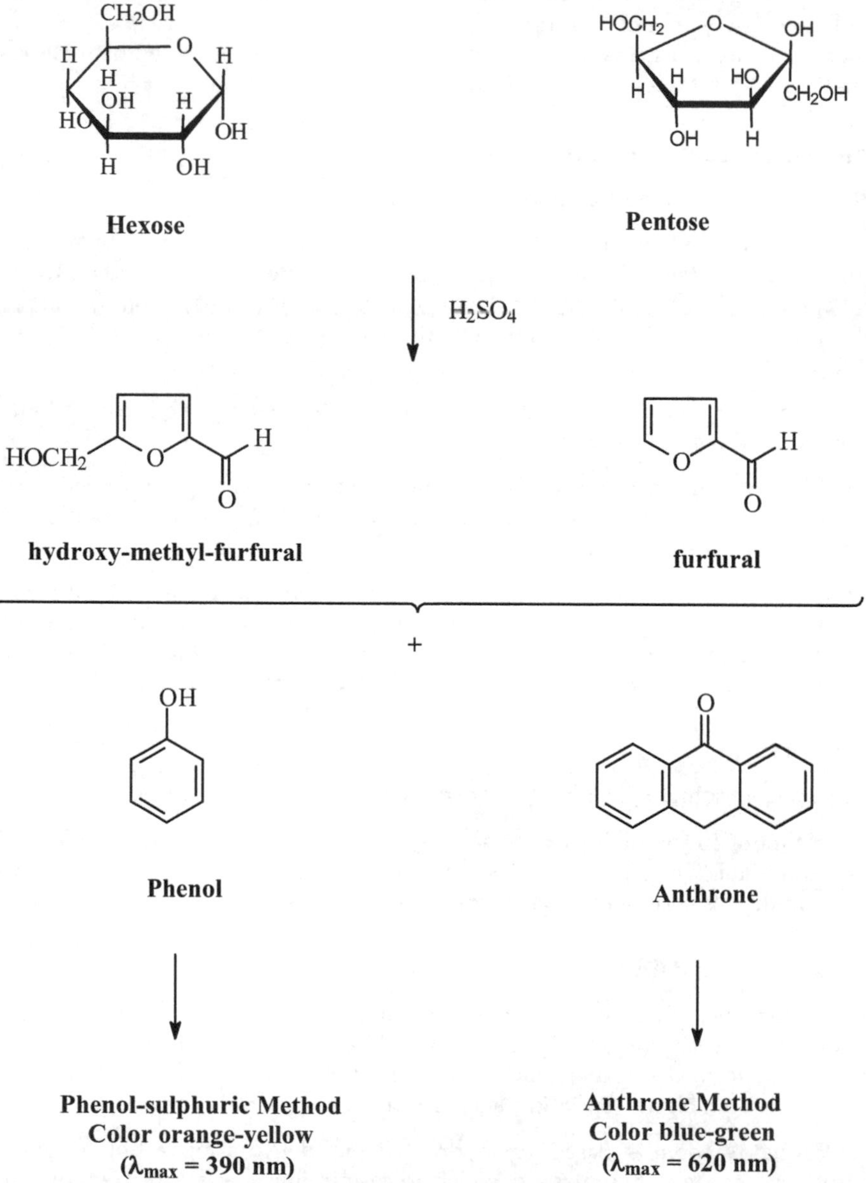

Fig. 1. Reaction of formation of the coloured complex in phenol-sulphuric and anthrone methods.

3.2.1 Methodology

Reagents: anthrone solution 0.2% (0.2g anthrone em q.s. to 100mL of sulphuric acid 95%). This solution must be kept at rest for 30 minutes with sporadic agitation until the solution to be clear, the reagents must be used in 12 hours.

Get an aliquot of 100µg.mL⁻¹ from the sample, add 2mL of the anthrone, leave at ice bath and hereupon cool in cooled water and read the absorbance in spectrophotometer at 620nm.

4. Reducing sugar determination

4.1 3,5-dinitrosalicylic acid method (Miller, 1959)

This method was first mentioned by Summer & Sisler (1944) and modified by Miller (1959), with this technique there is a possibility to dosage reducing sugar as total sugar.

The sugar act as a chemical reductor due free aldehyde group or ketone group presence in its molecule. In an alkaline medium, the reducing sugars are able to reduce the 3-5-dinitrosalicylic acid to 3-amino-5-nitrosalicylic acid, wherever, the aldehyde group is oxidized to aldonic acid (Figure 2). The 3-amino-5-nitrosalicylic acid is a orange color product, and the intensity of the color depends on the concentration of the reducing sugar. The sodium hydroxide provides the glucose reaction with 3-5-dinitrosalicylic acid by medium alkalinization.

Besides to 3-5-dinitrosalicylic acid, it is also used in this method the Rochelles salt (Potassium sodium tartrate), phenol, sodium bisulfite, and sodium hydroxide. The phenol optimize the quantity of the colour produced and sodium bisulfite stabilize the colour in the phenol presence (Miller, 1959).

The sensitivity of the method is from 100 to 500µg.mL⁻¹ of reducing sugar. A standard glucose or fructose solution is used to build the calibration curve and get the straight line equation to quantify samples.

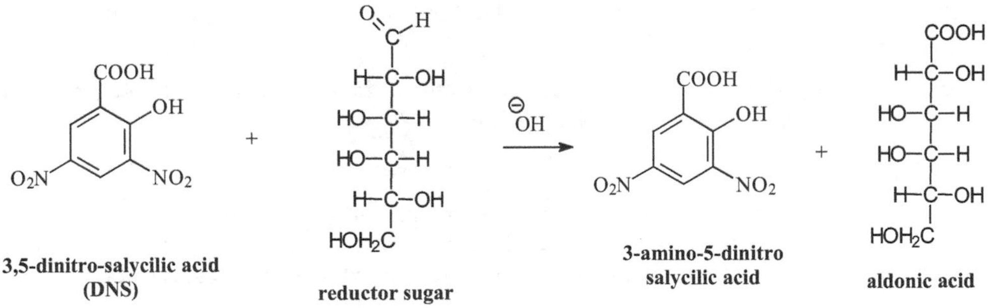

Fig. 2. Reaction of reducing sugar with 3,5-dinitro-salycilic acid reagent.

4.1.1 Methodology

Reagents:

Reagent A – Weight 2.5g DNS (3,5 dinitrosalicylic acid) and add 50mL NaOH 2M.

Reagent B – Weight 75g Potassium sodium tartrate (Rochelle salt), add 125mL of destiled water. Shake under heating until total dissolution.

Subsequently, add reagent A over the B, homogenize under heating until dissolve completely, and after cool the mix, complete the volume of the solution to 250mL.

Get an aliquot 500µg.mL^{-1} of the sample. Add 1mL of DNS, shake in vortex and let in a boiling bath during 5 minutes, put the material in ice bath until cool, add 3.75mL of destiled water, shake again, and read the absorbance in spectrophotometer at 540nm.

An adaptation of the DNS method to determine the total sugar can be getting from a previous hydrolysis before dosage. The hydrolysis is made with 0.5mL HCl concentrate and incubation in water bath at 60°C for 10 minutes. After then, the solution must be neutralized with NaOH 2M and cooled with ice bath until room temperature.

The procedure for total sugar quantification follow the same described for reducing sugar used above.

4.2 Hydrazide method of p-hydroxy-benzoic acid (Blakeney & Mutton, 1980)

This method is based on determination of reducing sugar before and after the digestion by invertase using acid for p-hydroxy benzoic acid hydrazide – PAHBAH. In this method, with some modifications, there is possible to perform the colorimetric determination of glucose, fructose and sucrose, during this analysis procedure with invertase. The major subject of the methodology described in this chapter is for determination of reducing sugar. The p-hydroxy benzoic acid hydrazide also has the advantage that glucose and fructose when reacting produce the same intensity of colour, and then all free monossacharides present in the sample can be determined.

4.2.1 Methodology

Reagents:

Reagent A: 10g p-hydroxy benzoic acid dissolved in 60mL water, add 10mL hydrochloride acid and complete volume to 200mL.

Reagent B: 24.9g trisodium citrate in 500mL water, 2.2g calcium chloride, 40g sodium hydroxide, and to complete the volume to 2.000mL.

In the day of analysis, mix reagents A and B in proportion of 1:10, after keep at 4°C.

Get an aliquot 50µg.mL^{-1} of the reducing sugar sample, add 5mL p-hydroxy benzoic acid hydrazide reagent (mixed in the same day) shake in vortex, incubate in water bath at 100°C by six minutes, cool until room temperature, and read the absorbance in spectrophotometer at 410nm.

5. Fructose determination

5.1 Sulphuric acid-Cysteine-Tryptophan method (Messineo & Edward Musarra,1972)

Fructose is dehydrated in acid medium for formation of furfural derivative, which complexes with cysteine hydrochloride to produce a chromophore at starting of green color that is unstable. Immediately, this first chromophore formed after react with tryptophan hydrochloride forming now a chromophore of pink colour that has greater stability chemical than the first and stability of 48 hours.

The reaction is also sensitive because can detect a low quantities like 1μg of fructose. The reaction also is specific because aldohexoses, aldopentoses, and ketopentoses do not interfere, however these compounds do not react even if its concentrations are higher, as 5mg.mL^{-1}. The sensitivity of the method is from 1 to 50μg of fructose. A standard solution of fructose is used to build the calibration curve and to get the equation of the linear regression to quantity samples.

5.1.1 Methodology

Reagents: sulphuric acid 75%, cysteine hydrochloride 2.5%, tryptophan solution in hydrochloride acid to formation of tryptophan hydrochloride (100μg.mL^{-1} in HCl 0.1M).

Getting an aliquot 50μg.mL^{-1} of the sample, add 2.8mL of sulphuric acid 75%, shake in vortex, add 0.1mL cysteine hydrochloride solution 2.5%, shake in vortex again, let in water-bath 45-50°C for 10 minutes, cool at room temperature and add 1mL tryptophan hydrochloride solution, shake again in vortex.

This sequence must be followed rigorously during assay because the formation of the final chromophore depends on the initial formation of the first formed chromophore by complexation with cysteine hydrochloride. After that, read the absorbance in spectrophotometer at 518nm.

5.2 Resorcinol method (Roe et al., 1949)

This reaction follows the same theoretical principles in which there is formation of furfural from hexoses and hydroxy-methyl-furfural (HMF) and from aldopentoses by acid dehydration (Fig. 1). These two products, singly are colorless, however, it is necessary a phenolic compound addition in the medium to develop a colored compound, in this case redness. This technique is firstly mentioned by Roe (1934), with some posterior modifications by Roe et al. (1949), becoming a quick reaction and with stable color. The reaction uses the hydrochloride acid (HCl) for carbohydrates dehydration and the resorcinol is the phenolic compound that reacts with furfural and HMF.

This test allows distinguish aldoses from ketoses because the reaction with ketoses is faster and more intense than aldoses. Therefore, the formation of the furfural is easier than HMF formation.

The sensitivity of the method ranges from 10 to 80μg.mL^{-1} fructose. A standard fructose solution is used to build the calibration curve and to get the equation of the linear regression to quantity samples.

5.2.1 Methodology

Reagents: Resorcinol reagent, 1g resorcinol and 0.25g thiourea in 100mL. This solution must be kept in the dark to keep its stability. Hydrochloride acid is diluted as 5mL HCl and 1mL water.

Getting an aliquot 80µg as maximum of the sample, add 0.5mL resorcinol reagent, shake in vortex, add 3.5mL hydrochloride acid, shake again, let in water bath at 80°C for 10 minutes, after that, read the absorbance in spectrophotometer at 520nm.

6. Enzymatic methods for sugar determination

There are several enzymatic methods for determination of the three principal sugars individually present in nectar - fructose, glucose, and sucrose, and in biological samples like plasma, blood, and urine. A lot of these methods are commercialized in kits and can be used successfully for rapid determination of the sugar from natural products samples. These kits are precise and sensitive, which enable rapid analysis and reliable results.

6.1 Glucose oxidase (Amaral et al., 2007)

The enzyme glucose oxidase is used for quantitative and enzymatic determination of the glucose in food and other materials. The enzyme glucose oxidase test is widely used because it is cheap, stable, and by its specificity well established for glucose.

In this reaction, the glucose is oxidized to gluconic acid and hydrogen peroxide by enzyme glucose oxidase. The hydrogen peroxide reacts with ortho-dianisidine in the presence of peroxidase enzyme to form a colored product (Figure 3). The compound of the ortho-dianisidine oxidation reacts with sulphuric acid to form a coloured product more stable. The intensity of pink colour measured in 540nm is proportional to the glucose concentration in the sample.

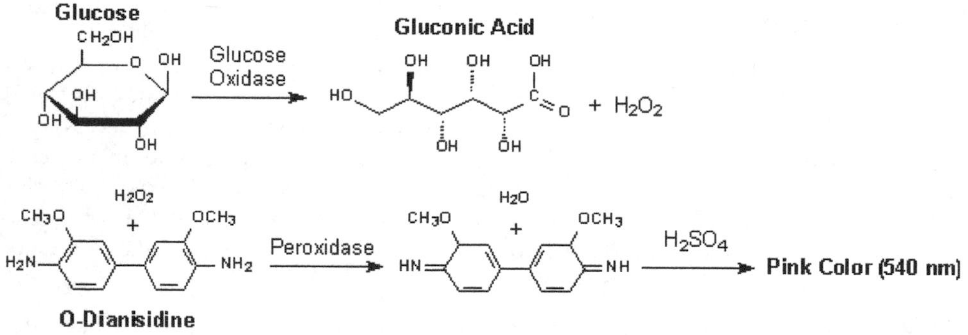

Fig. 3. Reactions of the enzymatic method of glucose oxidase.

6.2 Hexokinase method for simulteneous determination of glucose and fructose (Moerman et al., 2004)

This method is adequate for determination of monosaccharides glucose and fructose.

The principal enzyme of the method is hexokinase. This enzyme catalyzes the fosforilation of the glucose in glucose-6-phosphate, here upon the second enzyme the glucose-6-phosphate dehydrogenase together with cofactor nicotinamide adenine dinucleotide (NAD) oxidized glucose-6-phosphate to gluconate-6-phosphate and the NAD is reduced to NADH, according to stoichiometry of the second reaction (Figure 4), the spectrophotometric quantity of NADH is corresponding to glucose quantity.

Analysis of glucose is according to the following principle: hexokinase, a first enzyme, catalyzes the phosphorylation of glucose to glucose-6-phosphate, with the participation of the enzyme glucose-6-phosphate dehydrogenase and nicotinamide adenine dinucleotide (NAD) is further specifically oxidized to gluconate-6-phosphate. According to the stoichiometry of the last reaction (Figure 4), the photospectrometrically quantified amount of reduced nicotinamide adenine dinucleotide (NADH) is representative for the amount of glucose. Fructose is always determined subsequently to the glucose determination. Fructose undergoes phosphorylation to fructose-6-phosphate, with the same enzyme hexokinase, which is further converted to glucose-6-phosphate with phosphoglucose isomerase. Further oxidation to gluconate-6-phosphate as described above generates a supplementary amount of NADH that is stoichiometric with the amount of fructose.

All methodology of this analysis is performed following instructions of each enzymatic kit. For these enzymatic analysis must carried out the assays criteriously, because the order of addiction of reagents, the time of analysis and reading on spectrophotometer are determinants for an adequate analysis.

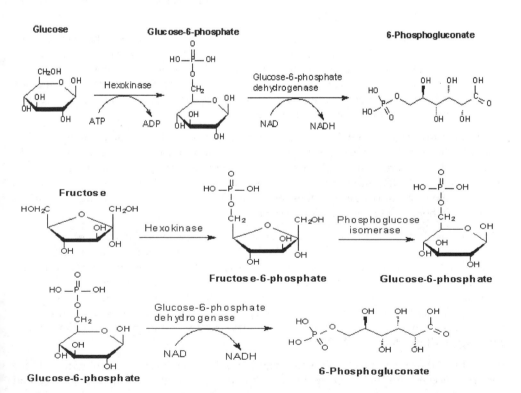

Fig. 4. Enzymatic reactions in enzymatic methods using hexokinase.

7. Sugar concentration in nectar

Nectar is considered the main reward to pollinator (Delaplane & Mayer, 2000) and its sugar concentration is associated to different pollinators, wherever the frequency and duration of visits depend on rate of nectar production (Biernaskie et al., 2002; Shafir et al., 2003; Nicolson & Nepi, 2005).

Toledo et al. (2005) reported total sugar concentration presents variation during the day in *Macroptilium atropurpureum* Urb., and can be related to number of visitor insects, specially, bees that collect nectar and pollen. In *Citrus sinensis*, it was verified that the high sugar concentration is an attractive, in special for *Apis mellifera*, and the availability of concentrated nectar during the day keeps the attractivity to pollinators (Malerbo-Souza et al., 2003). Therefore, the high quantity of nectar can leads to greater pollination rate by increasing in number of visitors bees (Silva & Dean, 2000).

Sugars are principal components of the floral nectar (Baker & Baker, 1973; Baker, 1977). The three most common sugar in nectar are sucrose, glucose and fructose in varying proportions (Freeman et al., 1985; Endress, 1994; Proctor et al., 1996; Baker et al., 1998). The amount of sugar secreted by flowers and consumed by pollinators cause a variation in sugar concentration of the flowers, during their anthesis period. Floral nectar consists of sugar pure solutions, specially glucose, sucrose and fructose (Roberts, 1979), however it can be found traces of oligosacchrides (Harbone, 1998).

Wykes (1952a) reported two oligosaccharides in nectar composition, the trisaccharide raffinose, and disaccharide melibiose in some nectaries. In some varieties of clover the presence of disaccharide maltose also was identified, however, this maltose can be a contaminant from aphids (Furgala et al., 1958). Taufel & Reiss (1952) confirmed that sucrose, glucose, and fructose, sugars promptly accepted by honeybees are current compounds present in nectar but another sugars can be present.

Besides, fructose and glucose, the presence of the monosaccharide D-galactose, in very low quantities (traces) also already was related in honey samples (Goldschmidt & Burket, 1955). However, it is important to emphasize that this monosaccharide when in its free form is considered a toxic compound to the honeybees (Siddiqui, 1970). Moreira & De Maria (2001) reviewed about carbohydrates in honey, and reported several di-, tri-, and oligosaccharides presented in honey and they came from nectar.

High-fructose is commonly used as sugar substitutes in processed foods, especially in soft drinks, mainly for economical reasons (Long, 1991). These products high-fructose corn syrups (HFCS) are obtained by enzymatic isomerization of corn syrups by both acid and enzymatic hydrolysis of cornstarch. Three enzymes are needed to transform cornstarch into the simple sugars glucose and fructose, α-amylase, glucoamylase, and glucose-isomerase. Fructosyl-fructoses were mainly detected in honey from honeybees fed with high-fructose corn syrups but not from those honeys coming from free-flying foragers or workers fed with sugar syrup (Ruiz-Matute et al., 2010).

Percival (1961) examined 889 plant species and found three pattern of carbohydrates to the nectar: a) nectar with high sucrose, nectar with similar quantities of sucrose, glucose and fructose; and nectar with high glucose and fructose. The nectar with sucrose dominant was associated to flowers of long tubes in which the nectar was protected (clovers), wherever the opened flowers had generally only glucose and fructose. These reports confirmed early researches (Wykes, 1953; Bailey et al., 1954) that suggested a relation between three monosaccharides and different species with flower. In another research, the nectar was divided in four different classes in function to sucrose/hexose rate - S/H: sucrose dominant - S/H >0.999, rich in sucrose - 0.5<0.999, rich in hexose - 0.1<0.499 and hexose dominant - S/H<0.1 (Baker & Baker, 1983).

In the research of Alves (2004) and Alves et al. (2010), the means of sucrose.hexose[-1] (S/H) per flower for all treatments were: $0.91\mu g.\mu L^{-1}$, for covered area with Africanized honeybee colony – rich in sucrose; $0.74\mu g.\mu L^{-1}$, semicovered area with free insects visitation – rich in sucrose; $0.86\mu g.\mu L^{-1}$, uncovered area with free insect visitation – rich in sucrose; and $3.05\mu g$ $.\mu L^{-1}$, for covered area without Africanized honeybee colony – sucrose dominant. However, Severson e Erickson (1984) reported in several cultivars of soybean values from 1.2:1.0:1.4 to 1.2:1.0:6.7, with sucrose predominance, which sucrose concentration in nectar ranged from 97 to $986\mu g.\mu L^{-1}$, these means are higher than those reported by Alves et al. (2010) who found $12.06\mu g.\mu L^{-1}$. This range suggests that sugar concentration in soybean nectar is influenced by other environment factors independently of pollinator action. Robacker et al. (1983) reported that edaphic and climatic factors affect the number of flowers and another floral characteristics during soybean growing. So, the environmental conditions that generate an increase in number and size of flowers, higher anthesis period, colourness more intense, and also greater nectar production are the factors responsible by became flowers more attractive to honeybees (Alves et al., 2010).

Cruden et al. (1983) suggested that the maximum nectar accumulation occurs before or at the beginning of pollination activity. Such fact can be verified in siratro, since the highest sugar concentration was found at 8:30 a.m. (Figure 5) time in which the bee visitation started (Toledo et al., 2005). Variations in the siratro nectar sugar content measured along the day were observed (Figure 6 – Toledo et al., 2005) and probably associated with the intensity of foraging by honeybees, which is directly related to the nectar quantity and quality (Heinrich, 1979; Hagler, 1990) or to its sugar composition (Waller, 1972; Abrol & Kapil, 1991).

In flowers exposed to pollinators, it is possible that nectar secretion ceases if there is not pollinator in the area or can be reabsorbed in old or pollinated flowers (Cruden et al., 1983). A nectar production without reabsorption may be have an impact on reproductive biology (Galleto & Bernardello, 1995). Therefore, plants reabsorb nectar from aging flowers and utilize its carbon in developing seeds and this is a reproductive advantage (Zimmerman, 1988).

Chiari et al. (2005) studying the pollination of Africanized honeybees on soybean flowers (*Glycine max* L. Merrill) var. BRS 133, measured through the manual refractometer the sugar concentration as total solids and concluded that data found presented a big uniformity, different of the results obtained by Sheppard et al. (1978) that observed big variations in these concentrations and attributed these differences to the variation in the soil composition and other environmental conditions, like precipitation. Despite this, the mean values found by Chiari et al. (2005) were 21.33 ± 0.22% in uncovered area and 22.33 ± 0.38% in covered area with honeybees and differed to each other (P=0.0001). Besides, the medium amounts of total sugar and glucose measured in the nectar of the flowers were 14.33 ± 0.96mg/flower and 3.61 ± 0.36mg/flower, respectively, in the same research.

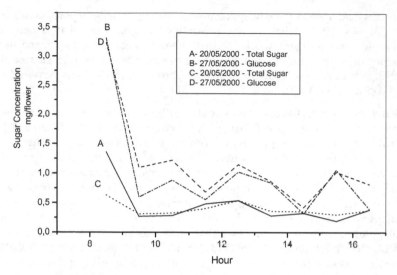

Fig. 5. Total sugar and glucose contents in nectar of siratro flowers along two-day period - reprinted from Toledo et al. (2005) with permission.

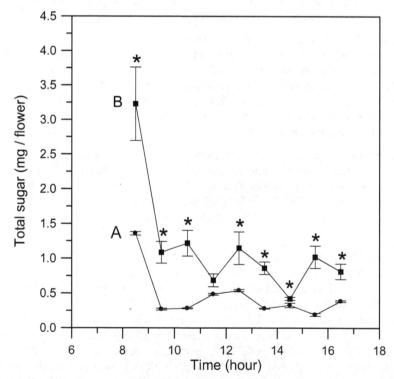

Fig. 6. Total sugar contents in nectar of siratro flowers along two-day period - reprinted from Toledo et al. (2005) with permission.

The study of 25 canola (*Brassica napus x Brassica campestris*) varieties carried out by Kevan et al. (1991) demonstrated that 23 of them had 0.95 or more glucose:fructose rates in their nectar. The same authors reported too that only three varieties had glucose in smaller quantities and none of the samples had detectable quantities of sucrose. Davis et al. (1998) reported higher glucose/fructose rate in lateral chambers than median in Brassicaceae.

Chromatography in paper showed that the nectar from siratro (*Macroptilium atropurpureum* Urb.) is constituted exclusively of glucose, as may be seen in Figure 5 (Toledo et al., 2005). It can be explain the low percentage of honeybee visit on siratro flower - 4%, wherever the another bees were *Trigona spinipes* - 24%, *Bombus morio* - 8%, *Euglossa sp* – 20%, *Megachilidae* - 12%, *Pseudaugochloropsis graminea* – 8% and *Halictidae* - 8% (Vieira et al., 2002). From this, it can be concluded that the main visitor and pollinator in siratro was *Euglossini*, however, *Trigona spinipes* perfurated the external part of the flower avoiding the contact with the pollen grains (Vieira et al., 2002).

Alves (2004) reported a variation of total sugar concentration in soybean nectar var. Codetec 207 from 21.33 to 27.47%, and the higher total sugar concentration was observed in covered area with Africanized honeybee colony. The sucrose concentration ranged from 9.63 to 13.61%, and the higher sucrose concentration was observed in covered area with Africanized honeybee colony. The glucose concentration in this variety was very low. The fructose concentration ranged from 7.93 to 13.75%, so the covered area with Africanized honeybee colony presented higher concentration too. Alves (2004) suggested that Africanized honeybees estimulated the sugar secretion in soybean nectar var. Codetec 207 (Table 1).

Aproximately, 20% of all food-crop production and about 15% of seed crops require the help of pollinators for full pollination (Klein et al., 2007), but Kevan & Phillips (2001) reported that aproximately 73% of cultivated vegetals in the world would be pollinated by some bee species. Gallai et al. (2009) reported a bioeconomic approach, which integrated the production dependence ratio on pollinators, for the 100 crops used directly for human food worldwide as listed by FAO. The total economic value of pollination worldwide amounted to €153 billion, which represented 9.5% of the value of the world agricultural production used for human food in 2005. The honeybee is the most common insect used as agricultural pollinator in many parts of the world like Europe and United States.

In United States, farmers rent more than 2 million honeybee colonies every year for pollination, but the honeybee is being threatened by several problems like Colony Colapse Desorder (James & Pitts-Singer, 2008). The cost of renting bees is up from US$50 per hive in 2003 to US$ 140 per hive in 2006 (Sumner & Boriss, 2006). Some crops require five to seven hives per hectare. In addition, no all crops are well pollinated by honeybees. For example, tomatoes require buzz pollination, which honeybees cannot achieve and alfalfa flowers are not properly worked by honeybees. Fortunately, honeybees are not the only bees that make good pollinators (James & Pitts-Singer, 2008).

Therefore, Greenleaf & Kremen (2006) reported that behavioral interactions between wild and honeybees increase the pollination efficiency of honeybees on hybrid sunflower up to 5-fold, effectively doubling honeybee pollination services on the average field. These indirect contributions caused interspecific interactions between wild and honeybees were more than five times more important than the contributions wild bees make to sunflower pollination directly.

Total Sugar	x±se (%)	n*	Rx	T2	T3	T4
			H=19.88		DF=3	KW=0.0002
covered area with honeybee colony	27.47±1.37	52	137.16	ns	0.0009	0.0008
semi-covered area	26.80±1.81	32	137.81		0.0039	0.0033
uncovered area	21.33±0.86	74	98.24			ns
covered area without honeybee	21.74±1.05	66	96.79			

Sucrose	x±se (%)	n	Rx	T2	T3	T4
			H=13.87		DF=3	KW=0.0031
covered area with honeybee colony	13.01±1.26	51	135.62	ns	0.0004	ns
semi-covered area	12.74±1.49	33	123.61		0.0028	ns
uncovered area	09.63±1.18	78	93.60			0.0366
covered area without honeybee	12.86±1.49	65	116.65			

Glucose	x±se (%)	n	Rx	T2	T3	T4
			H=0.78		DF=3	KW=0.8531
covered area with honeybee colony	0.71±0.05	81	135.62	ns	ns	ns
semi-covered area	0.78±0.09	52	123.61	ns	ns	ns
uncovered area	0.98±0.11	97	93.60	ns	ns	ns
covered area without honeybee	0.95±0.10	83	116.65	ns	ns	ns

Fructose	x±se (%)	n	Rx	T2	T3	T4
			H =12.26		DF=3	KW=0.0065
covered area with honeybee colony	13.75±1.36	51	59.30	ns	ns	0.0016
semi-covered area	13.29±1.38	33	58.42		0.0701	0.0075
uncovered area	10.72±1.05	78	46.33			0.0390
covered area without honeybee	7.93±1.32	65	34.92			

*n- sample size; x-Averages; se-standard error; H - H test; DF-degrees of freedom; KW-Kruskal-Wallis Probability; Rx-Medium position; ns-non significant and probability of the interactions (T1, T2, T3 and T4 vs T2, T3 and T4)

Table 1. Means of total sugar, sucrose, glucose and fructose concentration (%), in soybean nectar (Glycine max L. Merrill) var. Codetec 207 for covered area with honeybee colony (T1); Semi-covered area (T2); uncovered area (T3) and covered area without honeybee colony (T4) – Reprinted from Alves (2004) and Alves et al. (2010) with permission

Analysis of sugar composition in nectar can be used for detecting variation between flowers or nectaries from different taxonomic varieties, and consequently generate differences in type and frequency of visitation of pollinators. Alves et al. (2010) studied the total sugar concentration in soybean nectar (*Glycine max* L. Merrill) var. Codetec 207 by spectrophotometry, using the general method for carbohydrates determination by phenol-sulphuric technique (Dubois et al., 1956).

The research of Alves et al. (2010) was carried out in soybean plants in cages of $24m^2$ covered with an Africanized honeybee colony inside in which, semicovered area with free insect visits, uncovered area, covered area without Africanized honeybee colony. Each treatment was five repetitions. In this research, it was emphasized greater total sugar concentration in covered area with Africanized honeybee colony, reduction of sucrose concentration in uncovered area, and lower fructose concentration in covered area without honeybee colony and uncovered area. Besides, none difference among treatments in relation to glucose concentration. So, the high total sugar concentration observed in nectar of soybean var. Codetec 207, in covered area with Africanized honeybee colony suggests that the presence of *Apis mellifera* influenced in this composition, even though a variety with high grade of autopollination. However, the low fructose concentration in uncovered area can be related with low density of honeybees recorded or low presence of another preferential pollinator of fructose. For some researchers, the pollinators can affect the nectar composition (Canto et al., 2008; Herrera et al., 2009). In *Helleborus foetidus*, for example, some species of *Bombus* unchain modifications in sugar composition in nectar reducing the sucrose percentage, and rising the fructose and glucose percentage (Canto et al., 2008).

8. Conclusion

Nectar is an important floral reward for the bee visitation in the flowers. That food is converted in honey into the hive and used as an energy source for the workers. The bee-plant interaction is essential for the maintenance of genetic variability of plants as well as increased production of grains and fruits on a commercial scale.

Honeybees are very important for the pollination of both cultivated and native plants, and then understand the relationship between the collection of floral products (nectar, pollen, resins and oils), biology and behaviour of these insects help making better use for making bee products and agricultural products. Spectrophotometry is a tool to quantify and identify the components of floral products used by bees, especially nectar. These tests allow checking the correlation between the type and intensity of nectar in flower visitation by bees.

The methods employed in this molecular tool began to be developed in the early 50s of last century. Currently, the main analyses performed with the nectar spectrophotometry are: total sugars, reducing sugars, fructose determination (furfural and resorcinol), glucose oxidase (glucose determination in foods), hexokinase (glucose and fructose) and sugar concentration in nectar. As well as, in honey are made several analyses too, like diastase index, hydroxymethylfurfural and others.

Among the tests that can be done is determining sugar concentration in the nectar. The quantification of these sugars allows developing a series of studies associated with the floral visitation by bees collecting nectar and pollination of several species of cultivated plants. These studies contribute to the use of *A. mellifera* to assist beekeepers in increasing the honey production and the farmer in agricultural production. The association between agriculture and beekeeping has been demonstrated by studies presented that showed a positive association between the type of nectar produced by plant and intensity collection by honeybees. Much is still to be undertaken because the number of plant species studied that are pollinated by bees is limited in relation to the number of known species.

Especially in tropical regions, stingless native bees must be preserved and have great potential apicola and sustainable that is not operated, besides to the physicochemical analysis of the sugar composition of honey produced by these bees is not well known and do not have legislation around the world. Floral preferences of native stingless bees are not established yet. Spectrophotometry is an important tool to quantify and identify these components and contribute to knowledge about the interaction of stingless bees and flowers species in neotropical regions, besides to continue to be used in bee products chemical determination.

Finally it is worth mentioning that the spectrophotometry should be more standardized and published as an analysis tool to better use of the bees by the beekeepers in beekeeping industry and in agricultural production.

9. Acknowledgements

To the National Council for Scientific and Technological Development (CNPq), process number 479329/2009-5, and Coordination of Improvement Staff (CAPES) for their financial support.

10. References

Abrol, D.P. & Kapil, R.P. (1991). Foraging strategies of honeybees and solitary bees as determined by nectar sugar components. *Proceedings of the Indian National Academy of Sciences,* Vol. 57-B, p. 127-132, ISSN 0019-5588

Alves, E.M. (2004). *Polinização e composição de açúcares do néctar de soja (Glycine max L. Merrill) variedade Codetec 207.* Master dissertation, Universidade Estadual de Maringá, Maringá, Brasil. 57p.

Alves, E.M.; Toledo, V.A.A.; Oliveira, A.J.B.; Sereia, M.J.; Neves, C.A. & Ruvolo-Takasusuki, M.C.C. (2010). Influência de abelhas africanizadas na concentração de açúcares no nectar de soja *(Glycine max L. Merrill)* var. Codetec 207. *Acta Scientiarum – Animal Science,* Vol. 32, No. 2, pp. 189-195, ISSN 1807-8672

Amaral, L.I.V.; Gaspar, M.; Costa, P.M.F.; Aidar, M.P.M. & Buckeridge, M.S. (2007). Novo método enzimático rápido e sensível de extração e dosagem de amido em materiais vegetais. *Hoehnea,* Vol.34, p. 425-431, ISSN 0073-2877

Bailey, M.E.; Fieger, E.A. & Oertel, E. (1954). Paper chromatographic analyses of some southern nectars. *Gleanings in Bee Culture,* Vol. 82, p. 7-8, ISSN 0017-114X

Baker, H.G. & Hurd, Jr., P.D. (1968). Intrafloral ecology. *Annual Review of Entomology,* Vol.13, No.1, p. 385-414, ISSN 0066-4170

Baker, H.G. (1977). Non-sugar chemical constituents of nectar. *Apidologie,* Vol. 8, p. 349-356, ISSN 12979678

Baker, H.G. & Baker, I. (1973). Some anthecological aspects of the evolution of nectar producing flowers, particularly amino acid production in nectar. In: Heywood, V.H. (Ed.) *Taxonomy and ecology.* Academic Press, ISBN 0-12-346960-0, London, UK

Baker, H.G. & Baker, I. (1983). A brief historical review of the chemistry of floral nectar. In: Bentley, B. & Elias, T.S. (Eds.) *The biology of nectaries.* Columbia University Press, ISBN 978-0-231-04446-2, New York, USA

Baker, H.G.; Baker, I. & Hodges, S.A. (1998). Sugar composition of nectars and fruits consumed by birds and bats in the tropics and subtropics. *Biotropica,* Vol. 30, p. 559-586, ISSN 0006-3606

Biernaskie, J.M.; Cartar, R.V.; Hurly, T.A. (2002). Risk-averse inflorescence departure in hummingbirds and bumble bees: could plants benefit from variable nectar. *Oikos,* Vol. 98, No. 1, p. 98-104, ISSN 1600-0706

Blakeney, A.B. & Mutton, L.L. (1980). A simple colorimetric method for the determination of sugars in fruit and vegetables. *Journal of the Science of Food and Agriculture*, Vol.31, No.9, p. 889-897, ISSN 1097-0010

Buckmann, S.L. & Nabhan, G.P. (1997). *The forgotten pollinators*, Island Press, ISBN 1559633522, Washington, DC, USA

Canto, A.; Herrera, C.M.; Medrano, M.; Perez, R. & Garcia, I.M. (2008). Pollinator foraging modifies nectar sugar composition in *Helleborus foetidus* (Ranunculaceae): an experimental test. *American Journal of Botany*, Vol. 95, No. 3, pp. 315-320, ISSN 1537-2197

Chambó, E.D.; Garcia, R.C.; Oliveira, N.T.E. & Duarte, Jr., J.B. (2011). Honey bee visitation to sunflower: effects on pollination and plant genotype. *Scientia Agricola*, Vol.68, No.6, ISSN 0103-9016

Chiari, W.C.; Toledo, V.A.A.; Ruvolo-Takasusuki, M.C.C.; Attencia, V.M.; Costa, F.M.; Kotaka, C.S.; Sakaguti, E.S. & Magalhaes, H.R. (2005). Floral biology and behavior of Africanized honeybees *Apis mellifera* in soybean (*Glycine max* L. Merrill). *Brazilian Archives of Biology and Technology*, Vol. 48, No. 3, p.367-378, ISSN 1516-8913

Cruden, R.W.; Hermann, S.M. & Peterson, S. (1983). Patterns of nectar production and plant-pollinator coevolution. In: Bentley, B. & Elias, T. (Eds.). *The biology of nectaries*. Columbia, ISBN 023104461, New York, USA, p. 80-125

Delaplane, K.S.; Mayer, D.F. (2000). *Crop pollination by bees*. CABI Publishing, ISBN 9780851994482 London, UK

Dreywood, R. (1946). Qualitative test for carbohydrate materials. *Industrial and Engineering Chemistry Analytical Edition*, Vol.18, p. 499-504, ISSN 0096-4484

Dubois, M.; Giles, K.A. & Hamilton, J.K. (1956). Colorimetric method for determination of sugars and related substances. *Analytical Chemistry*, Vol. 28, p. 350-356, ISSN 0003-2700

Endress, P.K. (1994). *Diversity and evolutionary biology of tropical flowers*. Cambridge, ISBN 0-521-56510-3, Cambridge, USA

Free J.B. (1993). *Insect pollination of crops*. Academic Press, ISBN 0122666518, London, UK

Freeman, C.E.; Worthington, R.D. & Corral, R.D. (1985). Some floral nectar-sugar compositions from Durango and Sinaloa, Mexico. *Biotropica*, Vol. 17, p. 309-313, ISSN 0006-3606

Furgala, B.; Gochnauer, T.A. & Holdaway, F.G. (1958). Constituent sugars of some northern legume nectars. *Bee World, Vol. 39, p.* 203-205, ISSN 0005-772X

Galetto, L. & Bernardello, G. (1995). Characteristics of nectar secretion by *Lycium cestroides*, L. Ciliatum (Solanaceae), and their hybrid. *Plant Species Biology*, Vol. 11, p. 157-163, ISSN 1442-1984

Gallai, N.; Salles, J.-M; Settele, J. & Vaissière, B.E. (2009). Economic valuation of the vulnerability of world agriculture confronted with pollinator decline. *Ecological Economics*, Vol. 68, p. 810-821, ISSN 0921-8009

Goldschmidt, S. & Burkert, H. (1955). Observation of some heretofore unknown sugars in honey. *Zeitschrift für Physiologische Chemie*, Vol. 300, p. 188, ISSN 0018-4888

Greenleaf, S.S. & Kremen, C. (2006). Wild bees enhance honey bees'pollination of hybrid sunflower. *Proceedings of the National Academy of Sciences*, Vol. 103, No. 37, p. 13890-13895, ISSN 1091-6490

Hagler, J.R. (1990). Honeybee (*Apis mellifera* L.) response to simulated onion nectars containg variable sugar and potassium concentrations. *Apidologie*, Vol.21, p. 115-121, ISSN 12979678

Harbone, J.B. (1998). *Phytochemical methods: a guide to modern techniques of plant analysis*, Chapman & Hall, ISSN 0973-1296, New York, USA

Heinrich, B. (1979). Resource heterogeneity and patterns of movement in foraging bumblebees. *Oecologia*, Vol. 40, p. 235-245, ISSN 00298549

Herrera, C.M.; Veja, C.; Canto, A. & Pozo, M.I. (2009). Yeasts in floral nectar: a quantitative survey. *Annals of Botany*, Vol. 103, No. 9, pp. 1425-1423, ISSN 1095-8290

James, R.R. & Pitts-Singer, T.L. (2008). The future of agricultural pollination. In: _____ *Bee pollination in Agricultural Ecosystems*. Osford University Press, ISBN 978-0-19-531695-7, New York, USA

Kevan, P.; Lee, H. & Shuel, R.W. (1991). Sugar rations in nectars of varieties of canola (*Brassica napus*). *Journal of Apicultural Research*, Vol. 30, No. 2, p. 99-102, ISSN 0021-8839

Kevan, P. & Phillips, T.P. (2001). The economic impacts of pollinator declines: an approach to assessing the consequences. *Conservation Ecology*, Vol. 5, No.1, p. 1-13, available on line URL: http://www.consecol.org/vol5/iss1/art8/ accessed on Ap 14th 2009.

Klein, A.M.; Vaissière, B.E.; Cane, J.H. & Steffan-Dewenter, I. (2007). Importance of pollinators in changing landscapes for world crops. *Proceedings of the Royal Society of London: Series B*, Vol. 274, p. 303-313, ISSN 1471-2954

Lanza, J.S.; Smith G.C. & Sack, S. (1995). Variation in nectar volume and composition of *Impatiens capensis* at the individual, plant, and population levels. *Oecologia*, Vol.102, p. 113-119, ISSN 0029-8549

Long, J.E. (1991). High fructose corn syrup. *Food Science and Technology*, Vol. 48, 247-258, ISSN 1226-7708

Malagodi-Braga, K.S. (2005). Abelhas: por quê manejá-las para a polinização? *Revista Mensagem Doce*, No.80, ISSN 1981-6243

Malerbo-Souza, D.T.; Nogueira-Couto, R.H.; Couto L.A. (2003). Polinização em cultura de laranja (*Citrus sinensis* L. Osbeck, var. Pera-rio). *Brazilian Journal of Veterinary Research and Animal Science*, Vol. 40, No. 4, p. 237-242, ISSN 1413-9596

Malerbo-Souza, D.T.; Nogueira-Couto, R.H. & Couto, L.A. (2004). Honey bee attractants and pollination in sweet Orange, *Citrus sinensis* (L.) Osbeck, var. Pera-Rio. *Journal of Venomous Animals and Toxins including Tropical Diseases*, Vol.10, No.2, p. 144-153, ISSN 1678-9199

Martin, E.A.C.; Machado, R.J.P. & Lopes, J. (2005). Atrativo para abelhas em campos de produção de sementes de girassol colorido híbrido. *Semina: Ciências Agrárias*, Vol.26, No.4, p. 489-494, ISSN 1676-546X

Messineo, L. & Musarra, E. (1972). Sensitive spectrophotometric determination of fructose, sucrose, and inulin without interference from aldohexoses, aldopentoses, and ketopentoses. *International Journal of Biochemistry*, Vol.3, No.18, p. 691-699, ISSN 1357-2725

Moerman, F.T.; Van Leeuwen, M.B. & Delcour, J.A. (2004). Enrichment of higher molecular weight fractions in inulin. *Journal of Agricultural and Food Chemistry*, Vol.52, p. 3780-3783, ISSN 0021-8561

Moreira, R.F.A. & De Maria, C.A.B. (2001). Glicídeos no mel. *Química Nova*, Vol. 24, No. 4, p. 516-525, ISSN 1678-7064

Morgado, L.N.; Carvalho, C.F.; Souza, B. & Santana, M.P. (2002). Fauna de abelhas (Hymenoptera: Apoidea) nas flores de girassol *Helianthus annuus* L., em Lavras - MG. *Revista Ciência e Agrotecnologia*, Vol.26, No.6, p. 1167-1177, ISSN 1413-7054

Miller, G.L. (1959). Use of dinitrosalicylic acid reagent for determination of reducing sugar. *Analytical Chemistry*, Vol.31, No.3, p. 426-428, ISSN 0003-2700

Müller, A.; Diener, S.; Schnyder, S.; Stutz, K.; Sedivy, C.; Dorn, S. (2006). Quantitative pollen requirements of solitary bees: Implications for bee conservation and the evolution of bee-flower relationships. *Biological Conservation*, Vol.130, No.4, p. 604-615, ISSN 0006-3207

Nicolson, S.W.; Nepi, M. (2005). Dilute nectar in dry atmospheres: nectar secretion patterns in *Aloe castanea* (Asphodelaceae). *International Journal of Plant Sciences*, Vol. 166, No. 2, p. 227-233, ISSN 1058-5893

Paiva, G.J.; Terada, Y. & Toledo, V.A.A. (2002). Behavior of *Apis mellifera* L. Africanized honeybees in sunflower (*Helianthus annuus* L.) and evaluation of *Apis mellifera* L. colony inside covered area of sunflower. *Acta Scientiarum*, Vol.24, No.4, p. 851-855, ISSN 1679-9275

Percival, M. (1953). Types of nectar in angiosperms. *New Phytologist*, Vol. 60, p. 235-281, ISSN 1469-8137.

Pham-Delegue, M.H.; Etievant, P.; Ghuichard, E.; Marilleau, R.; Doualt, P.H.; Chauffaille, J. & Masson, C. (1990). Chemicals involved in honeybee sunflower relationship. *Journal of Chemical Ecology*, Vol.16, No.11, p. 3053-3065, ISSN 0098-0331

Prince, P.W. (1997). *Insect Ecology*, John Wiley & Sons, New York, ISBN 0471161845

Proctor, M.; Yeo, P. & Lack, A. (1996). *The natural history of pollination*, Harper Collins Publishers, London, ISBN 0002199068

Robacker, D.C.; Flottum, P.K. & Sammataro, D. (1983). Effects of climatic and edaphic factors on soybean flowers and of the subsequent attractiveness of the plants to honey bees. *Field Crops Research*, Vol. 6, No. 4, p. 267-278, ISSN 0378-4290

Roberts, R.B. (1979). Spectrophotometric analysis of sugars produced by plants and harvested by insects. *Journal of Apicultural Research*, Vol. 18, p. 191-195, ISSN 2078-6913

Roe, J.H. (1934). A colorimetric method for the determination of fructose in blood and urine. *Journal of Biological Chemistry*, Vol.107, p. 15-22, ISSN 0021-9258

Roe, J.H.; Epstein, J.H. & Goldstein, N.P. (1949). A photometric method for the determination os inulin in plasma and urine. *Journal of Biological Chemistry*, Vol.178, p. 839-845, ISSN 0021-9258

Ruiz-Matute, A.I.; Weiss, M.; Sammataro, D.; Finely, J. & Sanz, M.L. (2010). Carbohydrate composition of high-fructose corn syrups (HFCS) used for bee feeding: effect on honey composition. *Journal of Agricultural and Food Chemistry*, Vol. 58, p.7317-7322, ISSN 0021-8561

Schirmer, L.R. (1985). *Abelhas ecológicas*, Nobel, São Paulo: Nobel, ISBN 9253012536

Severson, D.W. & Erickson, J.E.H. (1984). Quantitative and qualitative variation in floral nectar of soybean cultivars in southeastern Missouri. *Environmental Entomology*, Vol. 13, No. 4, pp. 91-96, ISSN 1938-2936

Shafir, S.; Bechar, A.; Weber, E.U. (2003). Cognition mediated coevolution: context-dependent evaluations and sensitivity of pollinators to variability in nectar rewards. *Plant Systematic and Evolution*, Vol. 238, No. 1-4, p. 195-209, ISSN 1615-6110

Sheppard, W.S.; Jaycox, E.R. & Parise, S.G. (1978). Selection and management of honey bees for pollination of soybean. In: International Symposium of Pollination, 4., University of Maryland College Park, Maryland, USA. *Proceedings*... 1, p. 123-130

Siddiqui, I.R. (1970). The sugars of honey. *Advances in Carbohydrate Chemistry and Biochemistry*, Vol. 7, p. 51-59, ISSN 0065-2318

Silva, E.M.; Dean, B.B. (2000). Effect of nectar composition and nectar concentration on honey bee (Hymenoptera: Apidae) visitations to hybrid onion flowers. *Journal of Economic Entomology*, Vol. 93, No. 4, p. 1216-1221, ISSN 0022-0493

Sumner, D.A. & Boriss, H. (2006). Bee-conomics and the leap in pollination fees. *Giannini Foundation of Agricultural Economics*, Vol. 9, p. 9-11, ISSN 1081-6526

Summer, J.B. & Sisler, E.B. (1944). A simple method for blood sugar. *Archives of Biochemistry and Biophysics*, Vol.4, p. 333-336, ISSN 0003-9861

Täufel, K. & Reiss, R. (1952). Analytische und chromatographische Studien am Bienenhonig. *Zeitschrift für Lebensmittel-Untersuchung und Forschung*, Vol. 94, No. 1, p. 1-10, ISSN 0044-3026

Teixeira, L.M.R. & Zampieron, S.L.M. (2008). Estudo da fenologia, biologia floral do girasol (*Helianthus annuus*, Compositae) e visitantes florais associados, em diferentes estações do ano. *Ciência et Praxis*, Vol.1, No.1, p. 5-14, ISSN 1983192X

Thomson, J.D. & Goodell, K. (2002). Pollen removal and deposition by honeybee and bumblebee visitors to apple and almond flowers. *Journal of Applied Ecology*, Vol.38, No.5, p. 1032-1044, ISSN 365-2664

Toledo, V.A.A.; Oliveira, A.J.B.; Ruvolo-Takasusuki, M.C.C.; Mitsui, M.H.; Vieira, R.E.; Kotaka, C.S.; Chiari, W.C.; Gobbi Filho, L.; Terada, Y. (2005). Sugar content in nectar flowers of siratro (Macroptilium atropurpureum Urb.) *Acta Scientiarum – Animal Science*, Vol. 27, No. 1, p. 105-108, ISSN 1807-8672

Vidal, M.D.G.; De Jong, D.; Wien, H.C. & Morse, A.R. (2006). Nectar and pollen production in pumpkin (*Cucurbita pepo* L.). *Revista Brasileira de Botânica*, Vol.29, No.2, p. 267-273, ISSN 0100-8404

Vidal, M.D.G.; De Jong, D.; Wien, H.C. & Morse, A.R. (2010). Pollination and fruit set in pumpkin (*Cucurbita pepo*) by honey bees. *Revista Brasileira de Botânica*, Vol.33, No.1, p. 106-113, ISSN 0100-8404

Vieira, R.E.; Kotaka, C.S.; Mitsui, M.H.; Taniguchi, A.P.; Toledo, V.A.A.; Ruvolo-Takasusuki, M.C.C.; Terada, Y.; Sofia, S.H.; Costa, F.M. (2002). Biologia floral e polinização por abelhas em siratro (*Macroptilium atropurpureum* Urb.). *Acta Scientiarum*, Vol. 24, No. 4, p. 857-861, ISSN 1806-2636

Waller, G.D. (1972). Evaluating responses of honeybees to sugar solutions using an artificial-flower feeder. *Annals of the Entomological Society of America*, Vol. 65, p. 857-862, ISSN 0013-8746

Wykes, G. R. (1952a). An investigation of the sugars present in the nectar of flowers of various species. *New Phytologist*, Vol. 51, No. 2, p. 210-215, ISSN 1469-8137

Wykes, G.R. (1953). The sugar content of nectars. *Biochemical Journal*, Vol. 53, p. 294-296, ISSN 0264-6021

Yemm, E.W. & Willis, J. (1954). The estimation of carbohydrates in plant extracts by anthrone. *Journal of Biochemical*, Vol.57, No.3, p. 508-514, ISSN 0264-6021

Zimmerman, M. (1988). Nectar production, flowering phenology, and strategies for pollination. In: Doust, J.L. & Doust, L.L. (Eds.). *Plant reproductive ecology, patterns and strategies*. Oxford, ISBN 01-950-51750, New York, USA

2

Multivariate Data Processing in Spectrophotometric Analysis of Complex Chemical Systems

Zoltan Szabadai[2], Vicenţiu Vlaia[2], Ioan Ţăranu[1],
Bogdan-Ovidiu Ţăranu[1], Lavinia Vlaia[2] and Iuliana Popa[1]
[1]National Institute of Research-Development for Electrochemistry and Condensed Matter,
[2]University of Medicine and Pharmacy "Victor Babeş",
Romania

1. Introduction

There are a great variety of processing the analytical spectroscopy data, especially useful in multicomponent systems [Ewing et al., 1953; Garrido et al., 2004; Lykkesfeld, 2001; Oka et al., 1991; Sánchez & Kowalski, 1986]. These methods essentially are based on different strategies of mathematical strategies including specific formalism of mathematical statistics and of matrix algebra [Garrido et al., 2004; Szabadai, 2005]. The matrix-based methods reffer to quantitative analysis [Bosch-Reigh et al., 1991; Garrido et al., 2008; Li et al., 2011; Lozano et al., 2009; Ruckenbusch et al., 2006; Szabadai, 2005], to determination of the number of independent chemical equilibria in multicomponent systems [Szabadai, 2005] and for correction the action of various perturbing factors such as stray light or backgroud absorption [Burnius, 1959; Fox & Mueller, 1950; Melnick, 1952; Morton & Stubbs, 1946, 1947, 1948; Owen, 1995; Page & Berkovitz, 1943; Szabadai, 2005].

In the present chapter original approaches of matrix treatment of the aforementioned items are presented, with special consideration to the simultaneous assay of compounds in a mixter, to backgruond correction procedures and to the standard addition method in a generalized form.

2. Simultaneous assay of nonreacting compunds in a mixture

The issue of the quantitative analysis of a mixture, when the components do not interact chemically, can be approached, in a rigorous and general manner, with the help of matrix computation [Ewing et al., 1953; Garrido et al., 2004, 2008; Lozano et al., 2009; Lykkesfeld, 2001; Oka et al., 1991; Ruckenbusch et al., 2006; Sánchez & Kowalski, 1986; Szabadai, 2005]. In the case of a mixture with M component, the quantitative determination of the components, one has to measure the absorbance at Λ distinct values of wavelength ($\Lambda > M$). Given a set of N standard solutions ($N > M$ and supposing that, as a rule, each standard solution may contain all of M chemical components of interest in known concentrations), absorbances are to be measured at the same set of wavelengths and in identical conditions as done for standard solutions.

The following notations will be used in what follows: $X_m^n(\lambda)$ represents a quantity X referring to the standard mixture of number n (superscript index), at the individual chemical component of number m (subscript index), measured at the wavelength of number λ (between parentheses). Thus,

c_m^n represents the concentration of the component of number m in the standard solution of number "n" ;

c_m represents the concentration of the component of number m in the mixture undergoing the analysis (sample of unknown composition) ;

$A^n(\lambda)$ is the absorbance of the standard mixture of number m measured at the wavelength of number "λ" ;

$A_m(\lambda)$ is the contribution of the pure m-numbered component to the absorbance of the analyzed mixture, registered at the wavelength λ ;

$A(\lambda)$ is the absorbance of the mixture under analysis, measured at the wavelength of number λ ;

$\varepsilon_m(\lambda)$ is the molar absorptivity of the chemical component of number "m", measured at the wavelength of number λ ;

p^n is the weight percent of the spectrum of the standard solution of number n in the spectrum of the mixture under analysis.

If the components of a mixture do not interact chemically and if the absorbances of each component satisfies the Bouguer-Lambert-Beer relation, then the absorbance of the mixture, at each wavelength taken into account, consists of the sum of contributions of the individual absorbent chemical components. The absorbances of the N standard solutions, measured at Λ distinct values of wavelength, may be arranged in matrix form (1).

$$
\begin{bmatrix}
A^1(1) & \cdots & A^n(1) & \cdots & A^N(1) \\
\vdots & & \vdots & & \vdots \\
A^1(\lambda) & \cdots & A^n(\lambda) & \cdots & A^N(\lambda) \\
\vdots & & \vdots & & \vdots \\
A^1(\Lambda) & \cdots & A^n(\Lambda) & \cdots & A^N(\Lambda)
\end{bmatrix}
= d \cdot
\begin{bmatrix}
\varepsilon_1(1) & \cdots & \varepsilon_m(1) & \cdots & \varepsilon_M(1) \\
\vdots & & \vdots & & \vdots \\
\varepsilon_1(\lambda) & \cdots & \varepsilon_m(\lambda) & \cdots & \varepsilon_M(\lambda) \\
\vdots & & \vdots & & \vdots \\
\varepsilon_1(\Lambda) & \cdots & \varepsilon_m(\Lambda) & \cdots & \varepsilon_M(\Lambda)
\end{bmatrix}
\cdot
\begin{bmatrix}
c_1^1 & \cdots & c_1^n & \cdots & c_1^N \\
\vdots & & \vdots & & \vdots \\
c_m^1 & \cdots & c_m^n & \cdots & c_m^N \\
\vdots & & \vdots & & \vdots \\
c_M^1 & \cdots & c_M^n & \cdots & c_M^N
\end{bmatrix}
\quad (1)
$$

The left side of the relation includes the matrix of absorbances of the standard solutions and the optical path the radiation has been covered, „d" (i.e. the width of the cell used).

It may be allowed that the absorbance of the sample, measured at the same set of wavelengths as in the case of standard solutions, consists of the weighted contributions of the standard solutions. The contribution weight of each standard solution to the absorbance of the sample depends on the concentration of the chemical components in the sample under analysis and in the individual standard solutions. This is expressed, in matrix form, according to relation (2).

$$
\begin{bmatrix} A(1) \\ \vdots \\ A(\lambda) \\ \vdots \\ A(\Lambda) \end{bmatrix} = \begin{bmatrix} A^1(1) & \cdots & A^n(1) & \cdots & A^N(1) \\ \vdots & & \vdots & & \vdots \\ A^1(\lambda) & \cdots & A^n(\lambda) & \cdots & A^N(\lambda) \\ \vdots & & \vdots & & \vdots \\ A^1(\Lambda) & \cdots & A^n(\Lambda) & \cdots & A^N(\Lambda) \end{bmatrix} \cdot \begin{bmatrix} p^1 \\ \vdots \\ p^n \\ \vdots \\ p^N \end{bmatrix} \tag{2}
$$

In what follows, bold characters are used for denoting matrices: the matrix of the absorbances of the sample will be denoted by $\mathbf{A}$, the matrix of the absorbances of the standard solutions by $\mathbf{A_{st}}$, the matrix of the concentrations of chemical components in the analysed sample and in the standard solutions by $\mathbf{C}$ and $\mathbf{C_{st}}$ respectively, the matrix of the molar absorptivities by $\mathbf{E}$ and the matrix of the contribution weight of the standard solutions, generating the absorbance of the sample, by $\mathbf{P}$. In order to comprehend more easily the matrix formalism, the symbol of matrices is followed (between right brackets) by the specification of the number of rows and columns in the respective matrix. Therefore, matrix $\mathbf{A_{st}}$, made up of Λ rows and N columns, is denoted as follows : $\mathbf{A_{st}}[\Lambda,N]$. Relations (1) and (2) are equivalent to matrix expressions (3) and (4).

$$
\mathbf{A_{st}}[\Lambda,N] = d \cdot \mathbf{E}[\Lambda,M] \cdot \mathbf{C_{st}}[M,N] \tag{3}
$$

$$
\mathbf{A}[\Lambda,1] = \mathbf{A_{st}}[\Lambda,N] \cdot \mathbf{P}[N,1] = d \cdot \mathbf{E}[\Lambda,M] \cdot \mathbf{C}[M,1] \tag{4}
$$

Relations (1) and (2) may be written in a condensed matrix form (5).

$$
\frac{1}{d} \cdot \mathbf{A}[\Lambda,1] = \frac{1}{d} \cdot \mathbf{A}_{st}[\Lambda,N] \cdot \mathbf{P}[N,1] = \mathbf{E}[\Lambda,M] \cdot \mathbf{C}_{st}[M,N] \cdot \mathbf{P}[N,1] \tag{5}
$$

The product matrix $\mathbf{A_{st}}[\Lambda,N] \cdot \mathbf{P}[N,1]$ consisting of Λ rows and one column may be presented in the shortened form $(\mathbf{A_{st}} \cdot \mathbf{P})[\Lambda,1]$.

Practically, the aim is to calculate the elements of matrix $\mathbf{C}[M,1]$. In most of the real situations, the molar absorptivities of the chemical components under analysis are not known (especially not for a set of different wavelengths). For this reason, the spectrophotometric analysis is conditioned by the spectrophotometric study of a number of standard solutions, where the concentrations of the chemical components of interest are known. The matrix formalism presented allows for the standard solutions used to contain several chemical components (basically, each of the N standard solutions may contain all the M chemical components at known concentrations). In particular cases, it may happen (but it is not mandatory) that each standard solution contains only one chemical component (different from the other chemical components present in the other standard solutions); in this case the matrix $\mathbf{C_{st}}$ of the concentrations in standard solutions is square (has the same number of rows and columns) and diagonal (i.e. the c_m^n elements are null when m and n are different). In this particular case, the number of standard solutions is identical to the number of chemical components of analytical interest.

After the spectrophotometric measurements are accomplished, the elements of matrices $\mathbf{A_{st}}[\Lambda,N]$, $\mathbf{A}[\Lambda,1]$ and $\mathbf{C_{st}}[M,N]$ are known, and the further aim is to calculate the elements of matrix $C[M,1]$. These matrices satisfy relations (6) and (7). In what follows, the desired result is to eliminate matrix $E[\Lambda,M]$ from these two matrix relations and to explicit the resulting relation in relation to matrix $C[M,1]$.

$$A_{st}[\Lambda,N] = d \cdot E[\Lambda,M] \cdot C_{st}[M,N] \tag{6}$$

$$A[\Lambda,1] = d \cdot E[\Lambda,M] \cdot C[M,1] \tag{7}$$

In order to solve the above system of equation in relation to matrix $C[M,1]$, both members of equation (6) are multiplied on the right by the transpose of matrix $C_{st}[M,N]$.

$$A_{st}[\Lambda,N] \cdot C_{st}^{T}[N,M] = d \cdot E[\Lambda,M] \cdot C_{st}[M,N] \cdot C_{st}^{T}[N,M] \tag{8}$$

The product $C_{st}[M,N]\ C_{st}^{T}[N,M]$ is a MxM square matrix represented, according to the adopted notations, as $(C_{st} \cdot C_{st}^{T})\ [M,M]$. If the determinant of this matrix is not zero (i.e. if the set of wavelengths was selected suitably for relevant absorbance values), then the product matrix has an inverse, represented as $(C_{st} \cdot C_{st}^{T})^{-1}[M,M]$, with the property expressed by (9).

$$(C_{st} \cdot C_{st}^{T})^{-1}[M,M] \cdot (C_{st} \cdot C_{st}^{T})\ [M,M] = (C_{st} \cdot C_{st}^{T})\ [M,M] \cdot (C_{st} \cdot C_{st}^{T})^{-1}[M,M] = I[M,M] \tag{9}$$

In relation (9) $I[M,M]$ is the unit matrix of order M. The elements of this matrix situated on the main diagonal are equal to the unity, and all its other elements are null. The multiplication operation of any matrix by the unit matrix (of the corresponding order) leaves the matrix unchanged. Consequently, after multiplying the equation (8) on the right by $(C_{st} \cdot C_{st}^{T})^{-1}[M,M]$, the resulting relation is (10).

$$A_{st}[\Lambda,N] \cdot C_{st}^{T}[N,M] \cdot (C_{st} \cdot C_{st}^{T})^{-1}[M,M] = d \cdot E[\Lambda,M] \cdot I[M,M] = d \cdot E[\Lambda,M] \tag{10}$$

In what follows, both members of equation (7) are multiplied on the left by the transpose of matrix $E[\Lambda,M]$, namely by $E^{T}[M,\Lambda]$; the result is (11).

$$E^{T}[M,\Lambda] \cdot A[\Lambda,1] = d \cdot E^{T}[M,\Lambda] \cdot E[\Lambda,M] \cdot C[M,1] \tag{11}$$

The product $E^{T}[M,\Lambda] \cdot E[\Lambda,M] = (E^{T} \cdot E)[M,M]$ in the expression (11) is a square matrix allowing an inverse, $(E^{T} \cdot E)^{-1}[M,M]$, provided that the product matrix is not singular (its determinant is different fron zero). By multiplying equation (11) on the left by matrix $(E^{T} \cdot E)^{-1}[M,M]$, the expression (12) is obtained. This expresses explicitly the seeked column matrix $C[M,1]$ of the concentrations of components in the analysed mixture.

$$(E^{T} \cdot E)^{-1}[M,M] \cdot E^{T}[M,\Lambda] \cdot A[\Lambda,1] = d \cdot C[M,1] \tag{12}$$

Matrix $E[\Lambda,M]$, occuring in expression (4.43), can be calculated with relation (10).

In order to express the matrix of concentrations $C[M,1]$ only in relation to quantities resulting directly from spectrophotometric measurements (the elements of matrix $A[\Lambda,1]$) and in relation to known quantities (the elements of matrix $C_{et}[M,N]$), the matrix $E[\Lambda,M]$ has to be eliminated from relations (10) and (12).

The transpose of matrix $E[\Lambda,M]$, namely matrix $E^T[M,\Lambda]$, is expressed from relation (10) :

$$d \cdot E^T[M,\Lambda] = (C_{st} \cdot C_{st}^T)^{-1}[M,M] \cdot C_{st}[M,N] \cdot A_{st}^T[N,\Lambda] \qquad (13)$$

whereas the inverse matrix of the product of matrices $E[\Lambda,M]$ and $E^T[M,\Lambda]$ is expressed from (10) and (13):

$$(E^T E)^{-1}[M,M] = d^2 \cdot (C_{st} \cdot C_{st}^T)[M,M] \cdot (C_{st}[M,N] \cdot A_{st}^T[N,\Lambda] \cdot A_{st}[\Lambda,N] \cdot \qquad (14)$$

$$\cdot C_{st}^T[N,M])^{-1} \cdot (C_{st} \cdot C_{st}^T)^{-1}[M,M]$$

By replacing expressions (13) and (14) in (12), and taking into consideration relation (15),

$$(C_{st} \cdot C_{st}^T)^{-1}[M,M] \cdot (C_{st} \cdot C_{st}^T)[M,M] = I[M,M] \qquad (15)$$

the expression (16) is obtained. This presents, in an explicit form, the matrix of unknown concentrations.

$$C[M,1] = (C_{st} \cdot C_{st}^T)[M,M] \cdot (C_{st}[M,N] \cdot A_{st}^T[N,\Lambda] \cdot A_{st}[\Lambda,N] \cdot \qquad (16)$$

$$C_{st}^T[N,M])^{-1} \cdot C_{st}[M,N] \cdot A_{st}^T[N,\Lambda] \cdot A[\Lambda,1]$$

In relation (16) the optical pathway (d) no longer appears if the absorbances of standards $_{st}[\Lambda,N]$ and the absorbances of the sample $A[\Lambda,1]$ are measured at the same cell thickness.

Relation (10) allows to obtain the elements of matrix $E[\Lambda,M]$ as well, values which are proportional to the absorbances of the pure components measured at the selected Λ wavelengths. Relation (10) allows thus to obtain the spectrum of the M individual components. This is important if a sufficiently large number of standard solutions are available with known concentrations of components, but individual components are not available for recording their individual spectra.

A particular case of the above reasoning is that with each of standard solutions contain only one dissolved chemical component (other than those present in the other standard solutions), so N = M. In this case notation S refers to their common value (N = M = S). Consequently, matrix $C_{st}[S,S]$ of the concentrations of components in standard solutions is square and diagonal (only elements on the matrix main diagonal differ from zero) (17).

$$C_{st}[S,S] = \begin{bmatrix} c_1^1 & & & \\ & \ddots & & 0 \\ & & c_s^s & \\ 0 & & & \ddots \\ & & & & c_S^S \end{bmatrix} \qquad (17)$$

If the entry data (the absorbance readings at the selected wavelengths and the concentrations of the standard solutions) do not form sets of relevant data, then singular matrices may be obtained when processing the data (whose determinant is null), namely matrices which do not admit an inverse. In order to avoid this failure, the condition $\Lambda \geq N \geq M$ is imposed. This is the necessary (but no sufficient) condition to avoid the apparition of singular matrices. The necessity of the condition above results after inspecting the relations (10) and (12). In relation (10) the inverse of a matrix $(C_{st} \cdot C_{st}^T)^{-1}[M,M]$ appears calculated from matrix $C_{st}[M,N]$. Consequently, the matrix of the concentrations of the standard solutions must have higher – or at least equal – rank to the number M of chemical components in the sample. The necessary (but sufficient) condition for this requirement is $N \geq M$. In relation (12) the inverse of a matrix $(E^T \cdot E)^{-1}[M,M]$, is calculated from the matrix of molar absorptivities, $E[\Lambda,M]$. The necessary (but not sufficient) condition of the non-singularity of matrix $(E^T \cdot E)^{-1}[M,M]$ is the compliance of inegality $\Lambda \geq N$. The two necessary conditions for avoiding matrix singularity are expressed in the united form $\Lambda \geq N \geq M$. Also in order to avoid singularity in relation (10), the appropriate choice of concentrations of standard solutions is imposed, so that in matrix $C_{st}[M,N]$ both rows and columns should be linearly independent. Otherwise expressed, it is essential that there should not be any significant intercorrelation neither between different columns nor between different rows of the matrix of standard concentrations (in algebraic terms, the concentrations in standard solutions must form a complete basis in the linear M-dimensional field). In other words, the spectra of individual chemical components should differ significantly in the spectral field chosen for analysis (more precisely, for the selected set of wavelengths). The relevance of the choice of the wavelength set, from the point of view of the above-mentioned facts, can be tested by calculating the eigenvalues of the square and symmetric matrix $A_{st}^T[N,\Lambda] \cdot A_{st}[\Lambda,N]$. If one eigenvalue of this matrix is null (or very close to the null value), the selection of the wavelength set is not adequate for the intended analysis. The selection of another wavelength set is therefore necessary. The general issue of row (or column) intercorrelation is solved in linear algaebra by taking into consideration the issue of eigenvalues and eigenvectors. However, the complete and rigorous mathematical treatment of the issue of basis vectors in linear algaebra goes beyond the purpose of the present work.

2.1 Example

Let be $N = 5$ standard solutions containing $M = 3$ components of known concentrations. The concentrations, expressed in mg/l, are included in matrix $C_{st}[3,5]$. As illustrated by this matrix, each of the 5 standard solutions contains (in different and known concentrations) all three dissolved chemical components.

$$
C_{st}[3,5] = \begin{bmatrix} 2.50 & 4.25 & 1.25 & 0.85 & 2.22 \\ 3.00 & 1.00 & 1.62 & 1.15 & 3.36 \\ 4.00 & 0.80 & 5.00 & 4.45 & 0.82 \end{bmatrix} \; ; \; C_{st}^T[5,3] = \begin{bmatrix} 2.50 & 3.00 & 4.00 \\ 4.25 & 1.00 & 0.80 \\ 1.25 & 1.62 & 5.00 \\ 0.85 & 1.15 & 4.45 \\ 2.22 & 3.36 & 0.82 \end{bmatrix}
$$

The matrix $(C_{st} \cdot C_{st}^T)[3,3]$ resulting after multiplication and the eigenvalues of the product matrix ($EV[3,1]$) are illustrated below:

$$(C_{st} \cdot C_{st}^T)[3,3] = \begin{bmatrix} 31.5259 & 22.2117 & 25.2529 \\ 22.2117 & 25.2365 & 28.7727 \\ 25.2529 & 28.7727 & 62.1149 \end{bmatrix} \; ; \; _E V[3,1] = \begin{bmatrix} 4.915255 \\ 18.974022 \\ 94.988023 \end{bmatrix}$$

All three eigenvalues are different from zero (taking into account the concentration values and the precision in expressing concentration values), so the rank of the matrix $C_{st}[3,5]$ is 3. In other words, the set of concentration values allows to determine quantitatively all three chemical components in their mixture (provided that the wavelength set at which the absorbance values are going to be measured is chosen correctly).

The situation would differ if the matrix of concentrations of the standard solutions contained the following values:

$$C_{st}[3,5] = \begin{bmatrix} 2.50 & 4.25 & 1.25 & 0.85 & 2.22 \\ 3.00 & 1.00 & 1.62 & 1.15 & 3.36 \\ 5.50 & 5.25 & 2.87 & 2.00 & 5.58 \end{bmatrix} \; ; \; C_{st}^T[5,3] = \begin{bmatrix} 2.50 & 3.00 & 5.50 \\ 4.25 & 1.00 & 5.25 \\ 1.25 & 1.62 & 2.87 \\ 0.85 & 1.15 & 2.00 \\ 2.22 & 3.36 & 5.58 \end{bmatrix}$$

In this situation, the product $(C_{st} \cdot C_{st}^T)[3,3]$ of the two matrices has other eigenvalues.

$$(C_{st} \cdot C_{st}^T)[3,3] = \begin{bmatrix} 31.5259 & 22.2117 & 53.7376 \\ 22.2117 & 25.2365 & 47.4482 \\ 53.7376 & 47.4482 & 101.1858 \end{bmatrix}$$

$$_E V[3,1] = \begin{bmatrix} 5.966037 \\ 2.6393 \cdot 10^{-14} \\ 151.982163 \end{bmatrix}$$

In this case the rank of matrix $C_{st}[3,5]$ is only two because the second element in the column matrix of eigenvalues ($EV[3,1]$) is a lot smaller than the elements of the initial matrix and a lot smaller than the estimated accepted errors in expressing the standard concentrations. Consequently, even if a number of $N = 5$ standard solutions were used (with the considered concentrations), the concentrations of the three components in their mixture cannot be determined (irrespective of the wavelengths set chosen for measuring the absorbances), because the values of the concentrations of the standard solution have not been chosen properly.

2.2 Example

For numeric illustration of the spectrophotometric data processing with matrix formalism, the measurement data obtained analyzing the mixture of salicylic acid, caffeine and acetaminophen will be further presented [Szabadai, 2005]. The number of standard solutions is $N = 5$ and each standard solution contains all three components (in known concentrations). Table 1 contains absorbance values for the 5 standard solutions (A_{st}) and for the mixture of three substances (A), registered at the same set of 18 wavelengths. Table 1 also presents the known concentrations of the three components in the five standard solutions (elements of matrix $C_{st}[3,5]$), i.e. $M = 3$, $N = 5$, $\Lambda = 18$. The matrix of concentrations of the standard solutions $C_{st}[3,5]$, the matrix of absorbances of the standard solutions $A_{et}[18,5]$ and the matrix of absorbances of the sample $A[18,1]$ have the following forms:

Table 1

	Standard solution 1	Standard solution 2	Standard solution 3	Standard solution 4	Standard solution 5	Mixture (sample)
Absorbance values for different wavelengths (cell thickness d = 1 cm)	1.167	0.456	1.179	1.011	0.565	0.581
	1.192	0.435	1.257	1.048	0.513	0.566
	1.169	0.377	1.288	1.123	0.439	0.515
	1.109	0.290	1.265	1.109	0.374	0.443
	1.020	0.244	1.154	1.010	0.362	0.395
	0.932	0.228	1.000	0.867	0.402	0.370
	0.822	0.218	0.799	0.679	0.462	0.350
	0.747	0.217	0.645	0.534	0.524	0.336
	0.714	0.233	0.548	0.440	0.585	0.347
	0.654	0.232	0.487	0.388	0.552	0.329
	0.509	0.209	0.379	0.302	0.422	0.276
	0.295	0.167	0.228	0.183	0.225	0.195
	0.152	0.133	0.115	0.092	0.110	0.130
	0.089	0.100	0.060	0.046	0.069	0.095
	0.049	0.058	0.032	0.025	0.038	0.048
	0.030	0.026	0.023	0.018	0.021	0.030
	0.022	0.013	0.019	0.016	0.015	0.011
	0.020	0.009	0.018	0.015	0.013	0.010
Concentrations of components (mg/l) — Salicylic acid	2.50	4.25	1.25	0.85	2.22	
Caffeine	3.00	1.00	1.62	1.15	3.36	
Acetaminophen	4.00	0.80	5.00	4.45	0.82	

$$C_{st}[3,5] = \begin{bmatrix} 2.50 & 4.25 & 1.25 & 0.85 & 2.22 \\ 3.00 & 1.00 & 1.62 & 1.15 & 3.36 \\ 4.00 & 0.80 & 5.00 & 4.45 & 0.82 \end{bmatrix}$$

$$
A_{st}[18,5] = \begin{bmatrix}
1.167 & 0.456 & 1.179 & 1.011 & 0.565 \\
1.192 & 0.435 & 1.257 & 1.087 & 0.513 \\
1.169 & 0.377 & 1.288 & 1.123 & 0.439 \\
1.109 & 0.290 & 1.265 & 1.109 & 0.374 \\
1.020 & 0.244 & 1.154 & 1.010 & 0.362 \\
0.932 & 0.228 & 1.000 & 0.867 & 0.402 \\
0.822 & 0.218 & 0.799 & 0.679 & 0.462 \\
0.747 & 0.217 & 0.645 & 0.534 & 0.524 \\
0.714 & 0.233 & 0.548 & 0.440 & 0.585 \\
0.654 & 0.232 & 0.487 & 0.388 & 0.552 \\
0.509 & 0.209 & 0.379 & 0.302 & 0.422 \\
0.295 & 0.167 & 0.228 & 0.183 & 0.225 \\
0.152 & 0.133 & 0.115 & 0.092 & 0.110 \\
0.089 & 0.100 & 0.060 & 0.046 & 0.069 \\
0.049 & 0.058 & 0.032 & 0.025 & 0.038 \\
0.030 & 0.026 & 0.023 & 0.018 & 0.021 \\
0.022 & 0.013 & 0.019 & 0.016 & 0.015 \\
0.020 & 0.009 & 0.018 & 0.015 & 0.013
\end{bmatrix} ; \quad A[18,1] = \begin{bmatrix}
0.581 \\
0.566 \\
0.515 \\
0.443 \\
0.395 \\
0.370 \\
0.350 \\
0.336 \\
0.347 \\
0.329 \\
0.276 \\
0.195 \\
0.130 \\
0.095 \\
0.048 \\
0.030 \\
0.011 \\
0.010
\end{bmatrix}
$$

After performing the matrix operations in relation (16), the elements of matrix C[3,1] are obtained. They represent the concentrations, expressed in mg/l, of the three components of interest (salicylic acid, caffeine and paracetamol) in the analysed sample.

$$
C[3,1] = \begin{bmatrix}
3.538 \\
1.553 \\
1.381
\end{bmatrix}
$$

3. Generalization of the 3-point method to correct backgroud absorption

Before dealing generally with the issue of foreign components in the sample (components which cannot be found in standard solutions) – which may cause deviations from the hypothesis according to which the sample spectrum is formed by adding (with different weights) the spectra of standard solutions – the quantitative analysis method and the baseline correction algorithm suggested by *Morton* and *Stubbs* [Burnius, 1959; Ewing et al., 1953; Fox & Mueller, 1950; Melnick et al., 1952; Morton & Stubbs, 1946, 1947, 1948; Owen, 1995; Page & Berkovitz, 1943; Szabadai, 2005;] (also known as "3-point method") will be presented.

The *Morton* – *Stubbs* method takes into account that the sample often contains – besides the chemical substance of interest – other foreign absorbent chemical components. If the chemical removal of these foreign components is difficult, the elimination (or at least the minimisation) of their contribution to the final result of the analysis by correcting the absorbance read could be a comfortable solution. Accordind to the original form of the *Morton* and *Stubbs* method [Morton & Stubbs, 1946, 1947, 1948], it is possible to eliminate the disturbing effect of a foreign component only in the case in which the absorption of the disturbing component, manifested in the spectral field taken into consideration, does not present a maximum of absorption, but appears as a baseline absorption, dependent on the wavelength according to a linear function, which overlaps the absorption spectrum of the chemical component of interest.

The absorption spectrum of the component of intereset is deformed because of the background absorption (linearly dependent on the wavelength), and the effect of this deformation is eliminated through the special method of processing the measured absorbance values. According to the original *Morton – Stubbs* formalism, it is essential to determinate the absorbance of the sample at at least three wavelengths [Morton & Stubbs, 1946]. The wavelengths values involved are selected as follows: the wavelength used (λ_{max}) is the one at which the standard solution of the substance of interest (where the disturbing component is not present) presents a local absorbance maximum and another two wavelengths ((λ_1 and λ_2, λ_{max} being between these wavelengths) at which the substace of interest presents equal molar absorptivities ($\acute{A}(1) = \acute{A}(2)$). Figure 1 represents the spectrum of the standard solution by dotted line whereas the spectrum of the mixture, where the quantitative determination of the substances of interest is intended, is represented by a continuous line. The absorbance values corresponding to the three wavelengths selected (λ_1, λ_2 and λ_{max}) are denoted as A(1), A(2) and A(max) in the spectrum of the sample and as $\acute{A}(1)$, $\acute{A}(2)$ and $\acute{A}(max)$ in the spectrum of the pure (standard) component. The purpose is to calculate quantity $\acute{A}(max)$ (namely the absorbance associated with the substance of interest but without the backgroud absorbance) from the measured values A(1), A(2) and A(max). The absorbance $\acute{A}(max)$ is obtained by subtracting from the measured value A(max) the value denoted by x + y in Figure 1.

$$\acute{A}(max) = A(max) - (x + y) \tag{18}$$

The value x is expressed from the similarity of two triangles chosen conveniently:

$$\frac{\lambda_2 - \lambda_1}{\lambda_2 - \lambda_{max}} = \frac{A(1) - A(2)}{x}; \quad x = \frac{\lambda_2 - \lambda \, max}{\lambda_2 - \lambda_1} \cdot \left[A(1) - A(2) \right] \tag{19}$$

For calculating the value y in expression (18), the ratio of the absorbances $\acute{A}(max)$ and $\acute{A}(2)$ is needed, which can be determined from the spectrum of the standard solution. When elaborating an analytical method in order to determine a certain substance of interest, in a standardized work method, the ratio of the absorbances $\acute{A}(max)$ and $\acute{A}(2)$ once determined, it can be used for subsequent analyses, provided that analyses should be performed strictly in unchanged conditions (in the same solvent, at the same pH, the same temperature, with the same slit program of the spectrophotometer, preferably the same type of spectrophotometer as the one used for determining the above mentioned ratio). Let be denoted the aforementioned ratio as ρ:

$$\rho = \frac{A'(max)}{A'(2)} \tag{20}$$

In possession of the ratio ρ, the value y is obtained from relation (18) and (21).

$$\acute{A}(2) = A(2) - y \tag{21}$$

After dividing member by member relations (18) and (21), results:

$$\rho = \frac{A'(\max)}{A'(2)} = \frac{A(\max) - (x+y)}{A(2) - y} \; ; \; y = \frac{\rho \cdot A(2) - A(\max) + x}{\rho - 1} \tag{22}$$

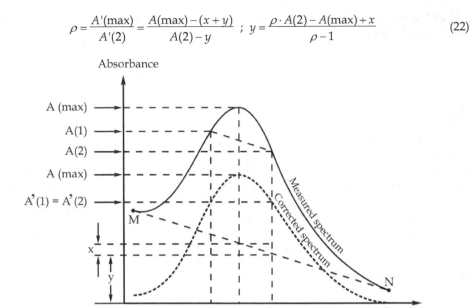

Fig. 1. Illustration of the *Morton – Stubbs* method

After replacing the expressions x and y in the latter relation, relation (23) results. It expresses the absorbance associated to the component of interest A′(max), which lacks baseline absorption.

$$A'(\max) = A(\max) - \frac{\lambda_2 - \lambda_{\max}}{\lambda_2 - \lambda_1} \cdot [A(1) - A(2)] -$$

$$- \frac{\rho \cdot A(2) - A(\max) + \frac{\lambda_2 - \lambda_{\max}}{\lambda_2 - \lambda_1} \cdot [A(1) - A(2)]}{\rho - 1} \tag{23}$$

The liniarity of the background absorption in a large spectral field is not always satisfied. In the case of wide absorption bands it is recommended to measure the absorbance of the sample at several wavelengths; in these cases however, the processing of the absorbance values measured requires more elaborated mathematical methods.

As it can be noticed, the *Morton – Stubbs* formalism allows the presence in the spectrum of the sample a linear background (a linear foreign spectrum in relation to the wavelength) which cannot be put down to any component of the standard solutions, ensuring corrected results (sample concentrations of the components of interest).

The original algorithm may be extended to ensure the obtention of corrected results in the case in which the sample spectrum contains, besides the chemical components represented in the standard spectra, a G degree polynomial baseline in relation to the wavelength. The spectrum of the sample is thus considered to consist of the spectra of the standard solutions and of the background spectrum, the latter being approximated to an adequate G degree polynomial (relation 24).

$$A(\lambda_i) = \sum_{k=1}^{K} p_k \cdot A_k(\lambda_i) + \sum_{g=0}^{G} q_g \cdot \lambda_i^g \quad ; \quad (i = 1, 2, ..., N) \tag{24}$$

The purpose is to calculate the contribution weight p_k of each standard solution to the spectrum of the sample, namely the coefficients p_k ($k = 1, 2, ..., K$). In the ideal case, when the spectrum of the sample does not contain a foreign baseline, but only the components represented in standard solutions, the coefficients q_g ($g = 0, 1, 2, ..., G$) are all null. Because of inherent measurement errors these coefficients are not null, but if the polynomial (25) is positive and has small values (for all wavelengths λ_i selected) in relation to the measured absorbances, the approach of the issue is correct and there are still chances to remove, by calculation, the effect of the polynomial backgroud (G degree) from the spectrum of the sample on the results. On the contrary, if the polynomial (25) has a high value or a negative one, even for one wavelength (one i value), the foreign backgroud cannot be approximated to a G degree polynomial form, and forcing the algorithm might lead to an erroneous result.

$$P(i) = \sum_{g=0}^{G} q_g \cdot \lambda_i^g \tag{25}$$

Obviously, the highest the G degree of the polynomial (25) which corrects the foreign backgroud in the spectrum of the sample, The more flexible the correction algorithm of a real backgroud absorption, but the more wavelengths should be selected where the absorbance readings are performed (in other words the inegality $N > G + K + 1$ is imposed in practice in order to obtain, from the measured absorbance values, a supra-determined system of equations).

For the statistical processing of the set of N absorbance values obtained for the sample and $N \times K$ absorbance values for the K standard solutions, the function (26) is defined imposing that for the values p_k ($k = 1, 2, ..., K$) and the values q_g ($g = 0, 1, ..., G$), which ensure the best global correspondence between the measured absorbances of the sample and the absorbances approximated with the relation (24), the function $F(p_k, q_g)$ should present a local minimum. The condition formulated is equivalent cancel the partial derivatives of the function (26) calculated in relation to p_k ($k = 1, 2, ..., K$) and q_g ($g = 0, 1, ..., G$). The cancellation of partial derivatives in (26) represents the necessary (but sufficient) condition for a local minimum of the function (26).

$$F(p_k, q_g) = \sum_{i=1}^{N} \left[A(\lambda_i) - \sum_{k=1}^{K} p_k \cdot A_k(\lambda_i) - \sum_{g=0}^{G} q_g \cdot \lambda^g \right]^2 = \min$$

$$\frac{\partial F(p_k, q_g)}{\partial q_{g^*}} = 0 \quad ; \quad \frac{\partial F(p_k, q_g)}{\partial p_{k^*}} = 0 \tag{26}$$

$$(k^* = 1, 2, ..., K \quad ; \quad g^* = 0, 1, 2, ..., G)$$

After derivation and equalization the derivatives to zero, a system of $K + G + 1$ linear equations is obtained, having the same number of unknowns (27).

$$\sum_{g=0}^{G} q_g \cdot \sum_{i=1}^{N} \lambda_i^g \cdot \lambda_i^{g^*} + \sum_{k=1}^{K} p_k \cdot \sum_{i=1}^{N} A_k(\lambda_i) \cdot \lambda_i^{g^*} = \sum_{i=1}^{N} A(\lambda_i) \cdot \lambda_i^{g^*}$$

$$(g^* = 0,1,2,\ldots,G)$$

$$\sum_{g=0}^{G} q_g \cdot \sum_{i=1}^{N} \lambda_i^g \cdot A_{k^*}(\lambda_i) + \sum_{k=1}^{K} p_k \cdot \sum_{i=1}^{N} A_k(\lambda_i) \cdot A_{k^*}(\lambda_i) = \sum_{i=1}^{N} A(\lambda_i) \cdot A_{k^*}(\lambda_i)$$

$$(k^* = 1,2,\ldots,K)$$

(27)

In order to express the wavelength (and its different powers) any unit of measure can be used, provided that the same unit of measure is used in all equations and for all wavelengths.

The generalisation of the *Morton – Stubbs* algorithm for the polynomial correction of the spectrum of the sample can also be presented in a matrix form. The equation system (24), written in a conventional algebraic form, is equivalent to matrix relation (28).

$$
\begin{bmatrix} A(\lambda_1) \\ A(\lambda_2) \\ \vdots \\ A(\lambda_i) \\ \vdots \\ A(\lambda_N) \end{bmatrix}
=
\begin{bmatrix}
A_1(\lambda_1) & \cdots & A_k(\lambda_1) & \cdots & A_K(\lambda_1) & 1 & \lambda_1 & \cdots & \lambda_1^g & \cdots & \lambda_1^G \\
A_1(\lambda_2) & \cdots & A_k(\lambda_2) & \cdots & A_K(\lambda_2) & 1 & \lambda_2 & \cdots & \lambda_2^g & \cdots & \lambda_2^G \\
\vdots & & \vdots & & \vdots & \vdots & \vdots & & \vdots & & \vdots \\
A_1(\lambda_i) & \cdots & A_k(\lambda_i) & \cdots & A_K(\lambda_i) & 1 & \lambda_i & \cdots & \lambda_i^g & \cdots & \lambda_i^G \\
\vdots & & \vdots & & \vdots & \vdots & \vdots & & \vdots & & \vdots \\
A_1(\lambda_N) & \cdots & A_k(\lambda_N) & \cdots & A_K(\lambda_N) & 1 & \lambda_N & \cdots & \lambda_N^g & \cdots & \lambda_N^G
\end{bmatrix}
\cdot
\begin{bmatrix} p_1 \\ \vdots \\ p_k \\ \vdots \\ p_K \\ q_0 \\ q_1 \\ \vdots \\ q_g \\ \vdots \\ q_G \end{bmatrix}
$$

(28)

If the matrix of the absorbance values of the sample (on the left member of the equation (28)) is denoted by $X[N,1]$, the first matrix factor on the right member by $Y[N,K+G+1]$ and the second matrix factor on the right member by $Z[K+G+1,1]$, the equation (28) can have the form (29).

$$X[N,1] = Y[N,K+G+1] \cdot Z[K+G+1,1] \tag{29}$$

The unknowns of interest are found in matrix $Z[K+G+1,1]$; the relative weights are p_k ($k = 1, 2, \ldots, K$). In order to explain the elements of matrix Z, both members of relation (29) are multiplied on the left by the transpose of matrix Y.

$$Y^T[K+G+1,N] \cdot X[N,1] = Y^T[K+G+1,N] \cdot Y[N,K+G+1] \cdot Z[K+G+1,1] \tag{30}$$

Matrix $(Y^T \cdot Y) = Y^T[K+G+1,N] \cdot Y[N,K+G+1]$ is square and allows an inverse matrix $(Y^T \cdot Y)^{-1}[N,N]$ if the associated determinant is not null. By multiplying relation (30) on the left by the inverse matrix, the explicit form of matrix Z results.

$$(\mathbf{Y^T \cdot Y})^{-1}[N,N] \cdot \mathbf{Y^T} \ [K+G+1,N] \cdot \mathbf{X}[N,1] = \mathbf{Z}[K+G+1,1] \tag{31}$$

It is decisively important to determine the correct set of wavelengths at which the absorbance values should be measured in case of a concrete analytical problem. The choice of the optimal wavelength (or wavelengths) is often a difficult issue even in case of a single component of interest. In real samples the component of interest may be accompanied by different other components whithout analytical interest ("the sample ballast"), but which can modify the molar absorptivity of the component of interest and so the sensitivity of the spectral answer of the chemical substance representing the object of the analysis. If one can identify the wavelength value at which the absorption of the sample ballast is negligible and at which the absorption of the component of interest is considerable, the respective wavelength is recommended for the determination. When at this wavelength the component of interest has even a local absorption maximum, this is an additional advantage, because at this wavelength the absorbance value depends in a minimum extent on the possible disorders in setting the wavelengths of the spectrophotometer. In the less fortunate case, where the sample ballast covers the entire spectral field available, more wavelengths are selected in order to determine the component in the sample in order to improve the specificity of the spectral answer in favour of the component of interest.

When the absorption of the component of interest and that of the ballast cannot be separated, a set of wavelengths can often be chosen so that the absorbances measured express the concentration of the component of interest through a multilinear relation (32).

$$c = f(\lambda_1) \cdot A(\lambda_1) + f(\lambda_2) \cdot A(\lambda_2) + \ldots + f(\lambda_i) \cdot A(\lambda_i) + \ldots + f(\lambda_N) \cdot A(\lambda_N) \tag{32}$$

The aim is to determine numerically the coefficients $f(\lambda_i)$; $(i = 1 , 2 , \ldots , N)$ for each wavelength in the spectral field considered (it is considered that the entire spectrum consists of N absorbance values associated to N discrete wavelength values) to calculate according to (32) the concentration of the chemical substance of interest in different samples, containing a different and unpredictable ballast. This purpose can sometimes be accomplished, sometimes not, according to the ballast variability of the analysed samples.

If there is any chance to determine a set of coefficients in agreement with the requirements mentioned for a component of interest, in presence of a ballast range in different samples, their calculation could be performed through calibration with a number of standard samples (let their number S) containing a ballast range as close as possible to that of real samples (of unknown composition) under analysis. Thus, two different standard samples may have the same concentration of the component of interest if they have a different ballast.

The concentrations of the component of interest in the S standard samples and the absorbances $A_s(\lambda_i)$; $(s = 1 , 2 , \ldots , S)$ of the standard samples satisfy the equation (32). The equation system obtained with the standard sample data can be rendered in matrix form (33).

If $S = N$, the number of unknowns equals the number of equations, so we dispose of the minimum number of equations necessary to solve the system (33) in relation to the N unknowns. For the reasons discussed above, the creation of a supra-determined system of equations is preferred $(S > N)$, as well as the search for a solution with an optimal global fit with least squares method.

$$
\begin{bmatrix} c_1^{st} \\ c_2^{st} \\ \vdots \\ c_s^{st} \\ \vdots \\ c_S^{st} \end{bmatrix} = \begin{bmatrix} A_1(\lambda_1) & A_1(\lambda_2) & \cdots & A_1(\lambda_i) & \cdots & A_1(\lambda_N) \\ A_2(\lambda_1) & A_2(\lambda_2) & \cdots & A_2(\lambda_i) & \cdots & A_2(\lambda_N) \\ \vdots & \vdots & & \vdots & & \vdots \\ A_s(\lambda_1) & A_s(\lambda_2) & \cdots & A_s(\lambda_i) & \cdots & A_s(\lambda_N) \\ \vdots & \vdots & & \vdots & & \vdots \\ A_S(\lambda_1) & A_S(\lambda_2) & \cdots & A_S(\lambda_i) & \cdots & A_S(\lambda_N) \end{bmatrix} \cdot \begin{bmatrix} f(\lambda_1) \\ f(\lambda_2) \\ \vdots \\ f(\lambda_i) \\ \vdots \\ f(\lambda_N) \end{bmatrix}
\tag{33}
$$

Values $f(\lambda_i)$; ($i = 1 , 2 , \dots , N$), representing the solution to the equation system (33), can be positive and negative numbers, both type being relevant for the analysis. If the absolute value of one of the coefficients $f(\lambda_i)$; ($i = 1 , 2 , \dots , N$) is small (negligible in relation to the mean of the absolute values of all coefficients), their contribution to the equation (32) is insignificant, they can be considered null, and the respective wavelengths are not relevant for the intended quantitative analysis. Therefore, to each wavelength in the spectrum a coefficient is associated expressing the relevance of that wavelength for the quantitative analysis of the component of interest in the presence of the matrix included in the calibration stage.

By excluding the irrelevant wavelengths, which do not improve the selectivity of the analytical method, one may reduce the number of wavelengths at which the measurement of absorbances is imposed when executing a real sample analysis.

In possession of the coefficients $f(\lambda_i)$; ($i = 1 , 2 , \dots , N$), the concentrations in the standard samples can be recalculated by relation (4.65) (the concentrations obtained are denoted c_1 , c_2 , ..., c_s , ..., c_S). Ideally, concentrations for all standard samples can be found. In reality, the correspondance between the set of existing (and known) concentrations in the S standard samples and the set of concentrations recalculated with relation (33) is not perfect. The success of the calibration operation can be expressed through the value of the linear correlation coefficient between the set of existing concentrations in the standard samples and the recalculated ones. Since the arithmetic mean of the existing (and known) concentrations in the S standard samples and the arithmetic mean of the concentrations recalculated with relation (33) are equal (according to a known theorem of mathematical statistics), their notation with a common symbol is justified:

$$
\overline{C} = \frac{1}{S} \cdot \sum_{s=1}^{S} c_s^{st} = \frac{1}{S} \cdot \sum_{s=1}^{S} c_s
$$

The linear correlation coefficient between the set of concentrations c_s^{st} and c_s ($s = 1 , 2 , \dots \dots$, S) is calculated with relation (34).

$$
r = \frac{\sum\limits_{s=1}^{S} \left(c_s^{st} - \overline{C} \right) \cdot \left(c_s - \overline{C} \right)}{\sqrt{\left[\sum\limits_{s=1}^{S} \left(c_s^{st} - \overline{C} \right)^2 \right] \cdot \left[\sum\limits_{s=1}^{S} \left(c_s - \overline{C} \right)^2 \right]}}
\tag{34}
$$

If the correlation coefficient (34) has an acceptable value from a statistical point of view (for example $r > 0{,}95$), it is likely that the set of coefficients $f(\lambda_i)$; ($i = 1 , 2 , \dots \dots , N$), obtained by solving the equation system (33) will allow to find the correct concentration of the substance of interest in real samples, provided that the real sample ballast is not completely different from the ballast range covered when calibrating the method (when determining the coefficients $f(\lambda_i)$; $(i = 1, 2, \ldots, N)$). This requirement is met to a certain extent in the case of serial analyses, where the nature of individual samples does not differ much, meaning that their ballast is similar.

Presenting a spectrum in a spectral field through pairs of wavelength-absorbance values $(A(\lambda)$ vs. λ , "digitized presentation") implies a large amount of data (for a faithful reprensentation of a spectrum the N number of sampling points is large). It results that, in order to genearate a supra-determinant equation system (33), an even larger number of standard samples is necessary $(S > N)$. This is generally inconvenient to realize in practice because it implies the use of a too large number of standard samples.

If $S < N$, the equation system (33) allows several sets of wavelengths for which the concentrations in standard samples correlate satisfactorily with the absorbance values, and the remaining problem is to identify at least one of these sets. This method is frequently used in practice, and establishing a profitable set of wavelengths involves the following stages:

(1) The matrix of absorbance values $A[S,N]$ turns into a new square matrix $B[S,S]$ whose columns are a complete orthogonal basis. The orthogonality of columns in the new matrix $B[S,S]$ can be realized, for example, by multiplying the matrix $A[S,N]$ on the right by a matrix $Q[N,S]$ chosen conveniently (35), so that the elements of matrix $B[S,S]$ = $A[S,N]\cdot Q[N,S]$ satisfy the orthogonality relation of columns (36). The construction of such a matrix $Q[N,S]$ is not unique; theoretically, there is an infinite number of such matrices capable of generating orthogonal columns satisfying the requirement (36). In spectrophotometric practice a diagonal-superior form of the matrix $Q[N,S]$ is sometimes used (where only elements on the main diagonal and those above this diagonal are different from zero).

$$\begin{bmatrix} A_1(\lambda_1) & \cdots & A_1(\lambda_i) & \cdots & A_1(\lambda_N) \\ \vdots & & \vdots & & \vdots \\ A_s(\lambda_1) & \cdots & A_s(\lambda_i) & \cdots & A_s(\lambda_N) \\ \vdots & & \vdots & & \vdots \\ A_S(\lambda_1) & \cdots & A_S(\lambda_i) & \cdots & A_S(\lambda_N) \end{bmatrix} \cdot \begin{bmatrix} Q_{11} & \cdots & Q_{1j} & \cdots & Q_{1S} \\ \vdots & & \vdots & & \vdots \\ Q_{j1} & \cdots & Q_{jj} & \cdots & Q_{jS} \\ \vdots & & \vdots & & \vdots \\ Q_{N1} & \cdots & Q_{Nj} & \cdots & Q_{NS} \end{bmatrix} = \begin{bmatrix} B_{11} & \cdots & B_{1j} & \cdots & B_{1S} \\ \vdots & & \vdots & & \vdots \\ B_{s1} & \cdots & B_{sj} & \cdots & B_{sS} \\ \vdots & & \vdots & & \vdots \\ B_{S1} & \cdots & B_{Sj} & \cdots & B_{SS} \end{bmatrix} \quad (35)$$

$$\sum_{s=1}^{S} B_s(\lambda_j) \cdot B_s(\lambda_{j^*}) = 0 \quad ; \quad \textit{for any } j \neq j^* \tag{36}$$

(2) Calculate the correlation coefficient of the elements of matrix $C_{et}[S,1]$ in relation (33) one by one with the columns of matrix $B[S,S]$ (for $j = 1, 2, \ldots, S$), thus obtaining N correlation coefficient values, in real cases all being smaller than theoretical value 1. The correlation coefficient of the elements of matrix $C_{st}[S,1]$ with the column "j" of matrix $B[S,S]$ is calculated by the relation (37).

$$r(\mathbf{C}_{st}[S,1], \mathbf{B}_j[S,1]) = \frac{\sum\limits_{s=1}^{S}\left(c_s^{s_i} - \overline{C}\right) \cdot \left(B_{sj} - \overline{B}_j\right)}{\sqrt{\left[\sum\limits_{s=1}^{S}\left(c_s^{st} - \overline{C}\right)^2\right] \cdot \left[\sum\limits_{s=1}^{S}\left(B_{sj} - \overline{B}_j\right)^2\right]}} \tag{37}$$

In relation (37) $\mathbf{B}_j[S,1]$ represents the column vector made up of the column of number "j" of matrix $\mathbf{B}[S,S]$, $\overline{C}$ is the mean value of the elements of matrix $\mathbf{C}_{st}[S,1]$ and $\overline{B}_j$ is the mean value of elements in column "j" in matrix $\mathbf{B}[S,S]$.

$$\overline{C} = \frac{1}{S} \cdot \sum_{s=1}^{S} c_s^{st} \quad ; \quad \overline{B}_j = \frac{1}{S} \cdot \sum_{s=1}^{S} B_{sj}$$

The correlation coefficients, calculated with relation (37) for j = 1, 2, . . . , S, in relation to "j" and the value "j" is retained (denoted by 1) for which the correlation coefficient is highest (in case of obtaining equal values of the correlation coefficient for more "j" values, one of these "j" values is retained arbitrarily).

(3) By using the multiple linear regression method, the elements of column matrix $\mathbf{C}_{et}[S,1]$ are correlated with all the pairs of columns of matrix $\mathbf{B}[S,S]$ obtained by combining column 1 with all the other columns of matrix $\mathbf{B}[S,S]$. The values of the multiple correlation coefficient $r(\mathbf{C}_{et}[S,1], (\mathbf{B}_1 \& \mathbf{B}_j)[S,2])$ are calculated in relation to the values taken by "j" and is retained (and denoted by 2) the value " j" for which the multiple correlation coefficient is highest. Two "j" values are thus obtained (denoted by 1 and 2) indicating the pair of columns in matrix $\mathbf{B}[S,S]$ which correlate conveniently with the column matrix $\mathbf{C}_{et}[S,1]$.

(4) By using the multiple linear regression method, the elements of column matrix $\mathbf{C}_{et}[S,1]$ are correlated with all sets of three columns of matrix $\mathbf{B}[S,S]$, obtained by combining columns 1 and 2 with all the other columns of matrix $\mathbf{B}[S,S]$. The values $r(\mathbf{C}_{et}[S,1], (\mathbf{B}_1 \& \mathbf{B}_2 \& \mathbf{B}_j)[S,3])$ are calculated in relation to the values taken by "j" and is retained (and denoted by 3) the value " j" for which the multiple correlation coefficient is highest. Three "j' values result this way (denoted by 1 , 2 , and 3), indicating the set of three columns of matrix $\mathbf{B}[S,S]$ which correlates conveniently with the column matrix $\mathbf{C}_{et}[S,1]$.

(5) The procedure described above continues by increasing progressively the number of columns of matrix $\mathbf{B}[S,S]$ with which is correlated, by multiple linear regression, the column matrix $\mathbf{C}_{et}[S,1]$. The columns of matrix $\mathbf{B}[S,S]$, involved at this phase, include those retained in the previous phase and a column which hasn not been yet retained. It is obvious that, by increasing the number of columns in $\mathbf{B}[S,S]$, involved in the multiple correlation, the optimal correlation coefficient approaches progressively the ideal value r = 1. Because the columns in matrix $\mathbf{B}[S,S]$ are orthogonal, there is no danger that, at a certain phase, the maximum correlation coefficient will be exceeded by a correlation coefficient corresponding to a combination of columns including a column (therefore a "j" value) which has not been retained in a previous phase. If the columns in matrix $\mathbf{B}[S,S]$ were not orthogonal, the above-mentioned danger would have appeared. This justifies the transformation of matrix $\mathbf{A}[S,N]$ (whose columns are not generally orthogonal) into a matrix $\mathbf{B}[S,S]$ with orthogonal columns. In practice, the procedure continues until obtaining a compromise situation, namely a satisfactory multiple correlation coefficient at a minimum number of involved columns if matrix $\mathbf{B}[S,S]$.

(6) Following the correlations described above, a set of columns of matrix $\mathbf{B}[S,S]$ results. These have been retained and denoted by $1, 2, \ldots, I$. A convenient set is then established, made up of wavelength values or, in other words, a set of I columns of the $\mathbf{A}[S,N]$ matrix. The set of I columns of matrix $\mathbf{A}[S,N]$ (conceived as I vectors in an imaginary S-dimensional space) is chosen so that each column of matrix $\mathbf{A}[S,N]$ presents a maximum covariance with a column of matrix $\mathbf{B}[S,S]$ retained during the above-mentioned operations. More concretely, if one suppose that the column of order "j" of matrix $\mathbf{B}[S,S]$ is associated to the column of order "i" of matrix $\mathbf{A}[S,N]$, it means that for the value "j" the column of order "i" of matrix $\mathbf{A}[S,N]$ ensures a maximum value of the covariance (of the correlation coefficient) calculated with relation (38).

$$r(\mathbf{A}_i[S,1], \mathbf{B}_i[S,1]) = \frac{\sum\limits_{s=1}^{S}(A_s(\lambda_i) - \overline{A}(\lambda_i)) \cdot \left(B_{sj} - \overline{B}_j\right)}{\sqrt{\left[\sum\limits_{s=1}^{S}(A_s(\lambda_i) - \overline{C})^2\right] \cdot \left[\sum\limits_{s=1}^{S}\left(B_{sj} - \overline{B}_j\right)^2\right]}} \tag{38}$$

In relation (38) $\overline{A}(\lambda_i)$ and $\overline{B}_j$ represent the arithmetic means of the corresponding matrix elements in columns of order "i", and "j" respectively.

$$\overline{A}(\lambda_i) = \frac{1}{S} \cdot \sum\limits_{s=1}^{S} A_s(\lambda_i) \quad ; \quad \overline{B}_j = \frac{1}{S} \cdot \sum\limits_{s=1}^{S} B_{sj}$$

In what follows, the wavelengths selected during phase (6) will be denoted by $\lambda_1^*, \lambda_2^*, \ldots, \lambda_j^*$. By applying relation (38) for $i = 1, 2, \ldots, N$ and $j = 1, 2, \ldots, I$, the matrix $\mathbf{R}[N,I]$ of the correlation coefficients is obtained (39).

$$\begin{bmatrix} r_{11} & \cdots & r_{1j} & \cdots & r_{1J} \\ \vdots & & \vdots & & \vdots \\ r_{i1} & \cdots & r_{ij} & \cdots & r_{iJ} \\ \vdots & & \vdots & & \vdots \\ r_{N1} & \cdots & r_{Nj} & \cdots & r_{NJ} \end{bmatrix} \tag{39}$$

In each column "j" of the matrix (39) an element r_{ij} with maximum absolute value is sought. The set of order numbers "i", which associates an "i" for each column"j", corresponds to the researched set of wavelengths.

(7) The equation system (15) is reconstructed, using only the set of wavelengths $\lambda_1^*, \lambda_2^*, \ldots, \lambda_j^*$ selected in previous phases.

$$\begin{bmatrix} c_1^{et} \\ c_2^{et} \\ \vdots \\ c_s^{et} \\ \vdots \\ c_S^{et} \end{bmatrix} = \begin{bmatrix} A_1(\lambda_1^*) & A_1(\lambda_2^*) & \cdots & A_1(\lambda_J^*) \\ A_2(\lambda_1^*) & A_2(\lambda_2^*) & \cdots & A_2(\lambda_J^*) \\ \vdots & \vdots & & \vdots \\ A_s(\lambda_1^*) & A_s(\lambda_2^*) & \cdots & A_s(\lambda_J^*) \\ \vdots & \vdots & & \vdots \\ A_S(\lambda_1^*) & A_S(\lambda_2^*) & \cdots & A_S(\lambda_J^*) \end{bmatrix} \cdot \begin{bmatrix} f(\lambda_1^*) \\ f(\lambda_2^*) \\ \vdots \\ f(\lambda_J^*) \end{bmatrix} \tag{40}$$

In order for the equation system (40) to be solvable in relation to the unknowns $f(\lambda_1^*)$, $f(\lambda_2^*)$, . . . , $f(\lambda_J^*)$, it is necessary that the number of selected wavelengths (J) be smaller than (or equal) to the number of standard samples (S). It is also essential that the determinant of matrix $D[J,J]$, resulting after multiplying the transpose of system matrix $(A^*)^T[J,S]$ by the system matrix $(A^*)[S,J]$ be significantly different from zero.

$$(A^{*T}[J,S] \cdot (A^*)[S,J] = D[J,J]; \det(D[J,J]) \neq 0$$

At the simultaneous determination of several chemical components which do not interact chemically, the equation system (1) and (2) has been constituted, with the help of N standard solutions, measured at Λ distinct wavelength values. In order to correctly solve the analytical problem, it is recommendable that the spectra of the N standard solutions be "as distinct as possible", because in the extreme (and imaginary) case where two standard solutions had identical spectra, the equation system would be undetermined, so impossible to solve. It is necessary to rigorously express the requirement that the spectra be as "different as possible". A method of characterizing the difference between spectra consists in considering the absorbances of a standard solution, measured at the selected set of wavelengths, as components of a vector in the Λ-dimensional space. The N spectra of standard solutions will thus form a set of N vectors.

$$D_{Gramm} = \begin{vmatrix} \sum_{\lambda=1}^{\Lambda} A^1(\lambda) \cdot A^1(\lambda) & \cdots & \sum_{\lambda=1}^{\Lambda} A^1(\lambda) \cdot A^n(\lambda) & \cdots & \sum_{\lambda=1}^{\Lambda} A^1(\lambda) \cdot A^N(\lambda) \\ \vdots & & \vdots & & \vdots \\ \sum_{\lambda=1}^{\Lambda} A^n(\lambda) \cdot A^1(\lambda) & \cdots & \sum_{\lambda=1}^{\Lambda} A^n(\lambda) \cdot A^n(\lambda) & \cdots & \sum_{\lambda=1}^{\Lambda} A^n(\lambda) \cdot A^N(\lambda) \\ \vdots & & \vdots & & \vdots \\ \sum_{\lambda=1}^{\Lambda} A^N(\lambda) \cdot A^1(\lambda) & \cdots & \sum_{\lambda=1}^{\Lambda} A^N(\lambda) \cdot A^n(\lambda) & \cdots & \sum_{\lambda=1}^{\Lambda} A^N(\lambda) \cdot A^N(\lambda) \end{vmatrix} \quad (41)$$

The value of the *Gramm* determinant (41) of the vector set expresses quantitatively the difference between vectors. The higher the value of the determinant (41), the more satisfied the requirement that the standard spectra be "as different as possible". At a higher value of the *Gramm* determinant the absorbance measurement error affects to a smaller extent the precision of the final results.

4. Generalization the standard addition method for several components of interest

In a real sample, subjected to be analyzed, one must take into consideration that the sample contains, besides the substance of interest, various other ingredients. Although it is possible to choose a wavelength at which the absorbance of the substance of interest should be significant and the absorbance of the ingredients negligible, it may happen that the ingredients, through their presence, modify the molar absorptivity of the component of interest, and thus modify the sensitivity of the spectrophotometric response to the component of interest. This possibility is more plausible in real pharmaceutical products, where the ingredients are found, as a rule, in a larger quantity than the active components. In this case, comparing the absorbance of the sample with that of a standard solution (which does not contain any ingredients) could provide erroneous analytical results. In order to realize even in these cases the quantitative determination of the active substance (the component of interest), one may resort to the "standard addition method" [Bosch-Reigh et al., 1991; Lozano et al., 2009 ; Szabadai, 2005; Valderrama & Poppi, 2009].

The reasoning of the addition method in the general case, when aiming to determine several components quantitatively, can be described with the help of the matrix calculation formalism [Szabadai, 2005]. The primary sample, in which the concentrations c_1 , c_2 , . . . , c_j , . . . , c_M of the M chemical components are analysed, is dissolved with an adequate solvent, bringing it to the final known volume V_a . A number of S + 1 equal portions (each having the volume "v") will be drawn from this solution. The portion number "0" is diluted to the final known volume V_b , thus obtaining the final solution of number "0" in which the concentrations of the components of interest are c_{10} , c_{20} , . . . , c_{M0}, and the concentration of ingredients is $c_{b(ing)}$. The portions number 1 , . . . , M are supplemented with known quantities of the M components of interest, so that, after completing to the final volume V_b, "S" solutions with modifications of known concentrations are obtained. In the final solution number "i", which was prepared by adding the masses m_{1i} , m_{2i} , . . . , m_{Mi} of individual components, the concentration modifications of components are Δc_{1i} , Δc_{2i} , . . . , Δc_{Mi} , whereas the concentration of ingredients remains the same in all S solutions, independent of "i". For each final solution the absorbance is measured at the same set of wavelengths λ_1 , λ_2 , . . . , λ_Λ. For the final solution number "0" the values $A_0(\lambda_1)$, $A_0(\lambda_2)$, . . . , $A_0(\lambda_L)$ are obtained. When measuring the absorbances of the final solutions of number 1 , 2 , . . . , S, at the same set of wavelengths and using the same optical path "d", the values $A_i(\lambda_1)$, $A_i(\lambda_2)$, . . . , $A_i(\lambda_L)$, i = 1 , 2 , . . . , S are obtained. The measured absorbances and the concentration modifications, generated by additions, can be arranged in matrix form. If $\varepsilon_j(\lambda)$ denotes the molar absorptivity of the component of order "j" at the wavelength "λ", the absorbances satisfy relations (42) and (43).

$$\frac{1}{d}\cdot\begin{bmatrix} A(\lambda_1) \\ \vdots \\ A(\lambda_\Lambda) \end{bmatrix} = \begin{bmatrix} \varepsilon_1(\lambda_1) & \cdots & \varepsilon_M(\lambda_1) \\ \vdots & & \vdots \\ \varepsilon_1(\lambda_\Lambda) & \cdots & \varepsilon_M(\lambda_\Lambda) \end{bmatrix}\cdot\begin{bmatrix} c_1 \\ \vdots \\ c_M \end{bmatrix} \tag{42}$$

$$\frac{1}{d}\cdot\begin{bmatrix} A^1(\lambda_1) & \cdots & A^S(\lambda_1) \\ \vdots & & \vdots \\ A^1(\lambda_\Lambda) & \cdots & A^S(\lambda_\Lambda) \end{bmatrix} = \begin{bmatrix} \varepsilon_1(\lambda_1) & \cdots & \varepsilon_M(\lambda_1) \\ \vdots & & \vdots \\ \varepsilon_1(\lambda_\Lambda) & \cdots & \varepsilon_M(\lambda_\Lambda) \end{bmatrix}\cdot\begin{bmatrix} c_1 + \Delta c_1^1 & \cdots & c_1 + \Delta c_1^S \\ \vdots & & \vdots \\ c_M + \Delta c_M^1 & \cdots & c_M + \Delta c_M^S \end{bmatrix} =$$

$$= \begin{bmatrix} \varepsilon_1(\lambda_1) & \cdots & \varepsilon_M(\lambda_1) \\ \vdots & & \vdots \\ \varepsilon_1(\lambda_\Lambda) & \cdots & \varepsilon_M(\lambda_\Lambda) \end{bmatrix}\cdot\left(\begin{bmatrix} c_1 & \cdots & c_1 \\ \vdots & & \vdots \\ c_M & \cdots & c_M \end{bmatrix} + \begin{bmatrix} \Delta c_1^1 & \cdots & \Delta c_1^S \\ \vdots & & \vdots \\ \Delta c_M^1 & \cdots & \Delta c_M^S \end{bmatrix}\right) = \tag{43}$$

$$= \frac{1}{d}\cdot\begin{bmatrix} A(\lambda_1) & \cdots & A(\lambda_1) \\ \vdots & & \vdots \\ A(\lambda_\Lambda) & \cdots & A(\lambda_\Lambda) \end{bmatrix} + \begin{bmatrix} \varepsilon_1(\lambda_1) & \cdots & \varepsilon_M(\lambda_1) \\ \vdots & & \vdots \\ \varepsilon_1(\lambda_\Lambda) & \cdots & \varepsilon_M(\lambda_\Lambda) \end{bmatrix}\begin{bmatrix} \Delta c_1^1 & \cdots & \Delta c_1^S \\ \vdots & & \vdots \\ \Delta c_M^1 & \cdots & \Delta c_M^S \end{bmatrix}$$

If the column matrix on the left member of the quation (42) is denoted by $A[\Lambda,1]$, the matrix of molar absorptivities on the right member of the equation (42) by $E[\Lambda,M]$ and the column matrix of the concentrations on the right member of the same equation by $C[M,1]$, the equation (42) takes the form (44).

$$(1/d)\cdot A[\Lambda,1] = E[\Lambda,M]\cdot C[M,1] \tag{44}$$

If equation (42) is subtracted, member by member, from equation (43) the result is equation (45).

$$\frac{1}{d}\cdot\left(\begin{bmatrix} A^1(\lambda_1) & \cdots & A^S(\lambda_1) \\ \vdots & & \vdots \\ A^1(\lambda_\Lambda) & \cdots & A^S(\lambda_\Lambda) \end{bmatrix} - \begin{bmatrix} A(\lambda_1) & \cdots & A(\lambda_1) \\ \vdots & & \vdots \\ A(\lambda_\Lambda) & \cdots & A(\lambda_\Lambda) \end{bmatrix}\right) =$$
$$= \begin{bmatrix} \varepsilon_1(\lambda_1) & \cdots & \varepsilon_M(\lambda_1) \\ \vdots & & \vdots \\ \varepsilon_1(\lambda_\Lambda) & \cdots & \varepsilon_M(\lambda_\Lambda) \end{bmatrix}\cdot\begin{bmatrix} \Delta c_1^1 & \cdots & \Delta c_1^S \\ \vdots & & \vdots \\ \Delta c_M^1 & \cdots & \Delta c_M^S \end{bmatrix} \tag{45}$$

$$\frac{1}{d}\cdot\begin{bmatrix} A^1(\lambda_1)-A(\lambda_1) & \cdots & A^S(\lambda_1)-A(\lambda_1) \\ \vdots & & \vdots \\ A^1(\lambda_\Lambda)-A(\lambda_\Lambda) & \cdots & A^S(\lambda_\Lambda)-A(\lambda_\Lambda) \end{bmatrix} =$$
$$= \begin{bmatrix} \varepsilon_1(\lambda_1) & \cdots & \varepsilon_M(\lambda_1) \\ \vdots & & \vdots \\ \varepsilon_1(\lambda_\Lambda) & \cdots & \varepsilon_M(\lambda_\Lambda) \end{bmatrix}\cdot\begin{bmatrix} \Delta c_1^1 & \cdots & \Delta c_1^S \\ \vdots & & \vdots \\ \Delta c_M^1 & \cdots & \Delta c_M^S \end{bmatrix}$$

Denoting by $\Delta A[\Lambda,S]$ the matrix of differences of absorbances in equation (45) and the matrix of concentration differences in (45), by $\Delta C[M,S]$, the resulting relation has the form (46).

$$(1/d)\cdot\Delta A[\Lambda,S] = E[\Lambda,M]\cdot\Delta C[M,S] \tag{46}$$

The matrix $E[\Lambda,S]$ is expressed from equation (46), and in its possession the equation (44) may be solved in relation to the column matrix $C[M,1]$. The necessary (but not sufficient) condition for solvency the equations in relation to matrix $C[M,1]$ is that Λ should be higher than (or equal) to M or S should be higher than (or equal) to M.

$$\Lambda \geq M \quad sau \quad S \geq M \tag{47}$$

In order to express the matrix $E[\Lambda,M]$, both sides of the relation (46) will be multiplied on the right by the transpose of matrix $\Delta C[M,S]$.

$$(1/d)\cdot\Delta A[\Lambda,S]\cdot\Delta C^{T}[S,M] = E[\Lambda,M]\cdot\Delta C[M,S]\cdot\Delta C^{T}[S,M] \tag{48}$$

Both sides of (48) are then multiplied by the inverse of matrix $\Delta C[M,S]\cdot\Delta C^{T}[M,S]$. Relation (49) is obtained, representing the explicit form of matrix $E[\Lambda,M]$.

$$(1/d) \cdot \Delta A[\Lambda,S] \cdot \Delta C^T[S,M] \cdot (\Delta C[M,S] \cdot \Delta C^T[S,M])^{-1} = E[\Lambda,M] \tag{49}$$

The concentration matrix $C[M,S]$ is expressed from relation (44). To this purpose, equation (44) is multiplied on the left by the transpose of matrix $E[\Lambda,M]$.

$$(1/d) \cdot E^T[M,\Lambda] \cdot A[\Lambda,1] = E^T[M,\Lambda] \cdot E[\Lambda,M] \cdot C[M,1] \tag{50}$$

When the above relation is multipled on the left by $(E^T[M,\Lambda] \cdot E[\Lambda,M])^{-1}$, the explicit form of the concentration matrix results (51).

$$(1/d) \cdot (E^T[M,\Lambda] \cdot E[\Lambda,M])^{-1} \cdot E^T[M,\Lambda] \cdot A[\Lambda,1] = C[M,1] \tag{51}$$

The particular case of standard addition method applied to a system with two components to be determined, is illustrated graphically in Figure 2. In this case, the procedure is reduced to determining the plane π passing through a number of figurative points and to reading the intersection points of this plane with the negative semi-axes of the concentrations.

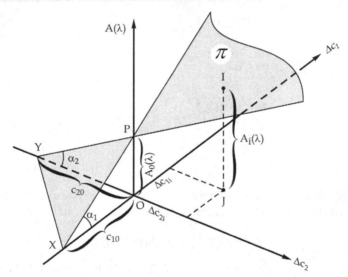

Fig. 2. Graphic representation of absorbances $A_i(\lambda)$ in relation to the modifications of concentrations Δc_{1i} and Δc_{2i} $(i = 1, 2, \ldots, n)$

At the graphic representation of absorbances $A_i(\lambda)$ vs. the increase of concentrations Δc_{1i} and Δc_{2i} $(i = 1, 2, \ldots, n)$, the figurative points are situated theoretically on a plane (denoted by π in Figure 4-20). The axis of absorbances is intersected by plane π in point P, corresponding to the absorbance $A_0(\lambda)$, measured in the case of the solution with $i = 0$. If at the selected wavelength (λ) the absorbance of the ingredients can be left out, the points X and Y, situated at the intersection of plane π with the negative parts of axes Δc_1 and Δc_2, have the coordinates $-c_{10}$ respectively $-c_{20}$ (in other words, the lengths of the segments OX and OY are proportional to the concentrations c_{10} and c_{20}). From the values c_{10} and c_{20}, and knowing the volumes V_a, v and V_b, one may calculate the concentrations c_1 and c_2 of the components of interest in the first solution, and finally their content in the primary sample.

4.1 Example

In order to illustrate the application of the standard addition method and of the subsequent data processing procedure, let consider the mixture of salicylic acid, caffeine and acetaminophen, discussed in a previous example. The aim is to determine the concentrations of the three chemical components. Table 2 includes the modifications of the component concentrations (5 modifications are performed) and the absorbances both for the original solution (where concentrations have not been modified) and for the five solutions in which the three chemical components have been modified. All absorbance values are read at the same set of 18 wavelengths ($\Lambda = 18$).

The elements of matrix E are calculated with relation (49) and are expressed in the tolerated unit of measure $l/(mol \cdot cm)$, employed in spectrophotometric practice, and the elements of matrix C, calculated with relation (51) are expressed in $\mu mol/l$.

Table 2		Modified solution 1	Modified solution 2	Modified solution 3	Modified solution 4	Modified solution 5	Original solution
		0.555	0.553	0.579	0.660	0.677	0.453
		0.519	0.525	0.552	0.627	0.637	0.424
		0.454	0.469	0.496	0.564	0.560	0.370
		0.382	0.405	0.430	0.498	0.471	0.311
		0.354	0.374	0.397	0.467	0.431	0.287
		0.368	0.378	0.398	0.476	0.439	0.297
		0.394	0.390	0.408	0.495	0.460	0.317
	Absorbance values at 18 wavelengths (d = 1 cm)	0.427	0.410	0.426	0.523	0.490	0.343
		0.466	0.440	0.455	0.560	0.531	0.374
		0.443	0.416	0.430	0.526	0.506	0.356
		0.350	0.329	0.340	0.409	0.405	0.283
		0.207	0.195	0.201	0.229	0.247	0.168
		0.119	0.111	0.115	0.120	0.149	0.098
		0.080	0.074	0.076	0.076	0.101	0.066
		0.045	0.041	0.043	0.042	0.057	0.037
		0.023	0.022	0.022	0.023	0.029	0.019
		0.014	0.014	0.014	0.016	0.018	0.012
		0.012	0.011	0.012	0.013	0.014	0.009
Δc_i $\mu mol/l$)	Salicylic acid	3.40	1.50	2.20	0.60	10.00	
	Caffeine	2.60	1.50	1.80	5.00	4.00	
	Acetaminophen	1.20	2.20	2.80	4.00	3.00	

$$
\mathbf{E} =
\begin{pmatrix}
6721.72 & 17534.60 & 28741.06 \\
5978.40 & 14156.50 & 32098.95 \\
4390.01 & 10405.40 & 34854.40 \\
1832.22 & 8685.62 & 35571.34 \\
675.67 & 10070.25 & 32288.01 \\
426.77 & 14575.14 & 26461.11 \\
463.41 & 20233.58 & 19124.73 \\
575.88 & 25871.08 & 12545.68 \\
1045.49 & 29980.13 & 8855.66 \\
1617.15 & 28153.68 & 7048.81 \\
2330.58 & 20505.79 & 5443.55 \\
3248.96 & 8960.11 & 3602.66 \\
3540.99 & 2713.00 & 1580.26 \\
2852.64 & 1204.43 & 613.63 \\
1694.67 & 550.32 & 343.30 \\
719.08 & 434.12 & 355.28 \\
292.02 & 443.65 & 350.29 \\
209.16 & 417.53 & 502.18
\end{pmatrix}
\quad ; \quad
\mathbf{C} =
\begin{pmatrix}
17.819 \\
10.290 \\
5.307
\end{pmatrix}
\mu mol / l
$$

5. Conclusions

The application of matrix algebra to the quantitative spectrophotometry provides a unified formalism for treatment the mathematical issues. Unlike the usual mathematical approaches, the matrix description of the phenomena behind the analytical spectrophotometry promise new dimensions for the automatic processing of results.

6. References

Bosch-Reigh, F., Campins-Falco, P., Sevillano-Cabeza, A., Herraes-Hernandez, R., & Molins-Legua C. (1991). Development of the H-Point Standard Addition Method for Ultraviolet-Visible Spectroscopic Kinetic Analysis of Two-component Systems, *Analytical Chemistry*, Vol.63, No.21, pp. 2424-2429

Burnius, E. (1959). Assay of Vitamin A Oils, *Journal of the American Oil Chemistry's Society*, Vol.35, No.4, pp. 13-14

Ewing, D.T., Sharpe L.H., & Bird O.D. (1953). Determination of Vitamin A in Presence of Tocopherols, *Analytical Chemistry*, Vol.25, No.4, pp. 599-604, ISSN 0003-2700

Fox, S.H. & Mueller, A. (1950). The Influence of Tocopherols on U.S.P. XIV Vitamin Assay, *Journal of the American Pharmaceutical Association*, Vol.39, No.11, pp.621-623

Garrido, M., Lázaro, I., Larrechi, M.S., & Rius F.X. (2004). Multivariate Resolution of Rank-deficient Near-infrared Spectroscopy Data fron the Reaction of Curing Epoxi Resins Using the Rank Augmentation Strategy and Multivariate Curve Resolution Alternating Least Squares Approach, *Analytica Chimica Acta*, Vol.515. No.1, pp.65-73

Garrido, M., Rius, F.X., & Larrechi, M.S. (2008). Multivariate Curve Resolution-alternating Least Squares (MCR-ALS) Applied to Spectroscopic Data from Monitoring Chemical Reaction Processes, *Anal. Bioanal. Chem.*, Vol.390, No.8, pp. 2059-2066

Li, N., Li, X.Y., Zou, Z.X., Lin, L.R., & Li, Y.Q. (2011). A Novel Baseline-correction Method for Standard Addition Based Derivative Spectra and its Applications to Quantitative Analysis of benzo(a)pyrene in Vegetable Oil Samples, *Analyst*, Vol.136, pp. 2802-2810

Lozano, V.A., Ibañez, G.A., & Olivieri, A.C. (2009). A Novel Second-order Standard Addition Analytical Method Based on Data Processing with Multidimensional Partial Least-squares and Residual Bilinearization, *Analytica Chimica Acta*, Vol.651, No.2, pp. 165-172

Lykkesfeld, J. (2001). Determination of Malonaldehyde as Dithiobarbituric Acid Adduct in Biological Samples by HPLC with Fluorescence Detection: Comparison with Ultraviolet-Visible Spectrophotometry, *Clinical Chemistry*, Vol.47, pp.1725-1727

Melnick, D., Luckmann, F.H., & Vahlteich, H.W. (1952). Estimation of Vitamin A in Margarine. I. Collaborative Study of Assay Methods for Estimating the Potencu of the Vitamin A Concentrates, *Journal of the American Oil Chemistry's Society*, Vol.29, No.3, pp. 104-108

Morton, R.A., & Stubbs A.L., (1946). *Photoelectric Spectrophotometry Applied to the Analysis of Mixtures and Vitamin A Oils*, Vol.71, pp. 348-350

Morton, R.A., & Stubbs, A.L. (1947). A Re-examination of Halibut-liver Oil. Relation Between Biological Potency and Ultraviolet Absorption Due to Vitamin A, *Biochem. J.*, Vol.41, pp. 525-529

Morton, R.A., & Stubbs, A.L. (1948). Studies in Vitamin A. 4. Spectrophotometric Determination of Vitamin A in Liver Oils. Correction for Irrelevant Absorption, *Biochem. J.*, Vol.42, No.2, pp. 195-203

Oka, K., Oshima, K., Inamoto, N., & Pishva D. (1991). Chemometrics and Spectroscopy, *Analytical Sciences*, Vol.7, pp.757-760, ISSN 0910-6340

Owen, A.J. (1995). Quantitative UV-Visible Analysis in the Presence of Scattering" Application Notes – Agilent Technologies, *Pharmaceutical Analysis*, (publication number: 5963-3937E)

Page, R.C., & Berkovitz, Z. (1943). The Absorption of Vitamin A in Chronic Ulcerative Colitis, *American Journal of Digestive Diseases*, Vol.10, No.5, pp. 174-177

Ruckenbusch, C., De Juan, A., Duponchel, L., & Huvenne, J.P. (2006). Matrix Augmentation for Breaking Rank-deficiency: A Case Study, *Chemometrics and Intelligent Laboratory Systems*, Vol.80, No.2, pp. 209-214

Sánchez, E.; & Kowalski, B.R. (1986). Generalized Rank Annihilation Factor Analysis, *Analytical Chemistry*, Vol.58, pp. 496-499, ISSN 0003-2700

Szabadai Z. (2005). *Bazele fizico-chimice ale metodelor de control analitic al medicamentelor*, Editura Mirton, Vol. II., pp. 4.38-4.96, ISBN: 973-661-677-0, Timişoara, Romania

Valderrama, P., & Poppi R.J., (2009). Second Order Standard Addition Method and
    Fluorescence Spectroscopy in the Quantification of Ibuprofen Enantiomers in
    Biological Fluids, *Chemometrics and Intelligent Laboratory Systems*, Vol.106, No.2, pp.
    160-165.

3

Basic Principles
and Analytical Application
of Derivative Spectrophotometry

Joanna Karpinska

Institute of Chemistry,
University of Bialystok, Bialystok
Poland

1. Introduction

Analytical methods based on measurements of UV or visible light absorption belong to the most popular and most often used in laboratory practice. Commercially available apparatuses are cheap and easy for operation. Spectrophotometric procedures usually are not time- and labour-consuming. The economical aspects of UV-Vis techniques is worth of emphasize too. It is one of the cheapest technique, so spectrophotometers are basic equipment of every laboratory. The main disadvantage and limitation of the spectrophotometry is its low selectivity. The measurement of absorbance is burden by interferences derived from others components of sample. A recorded UV-Vis spectrum is the sum of absorbances of analyte and matrix. Usually, recorded bands are well-defined but more or less distorted by a background. As the background is called absorbance exhibited by matrix (reagents or accompanied compounds). The problem with specific or nonspecific background can be omitted by measurements versus blank. Such procedure can be successfully applied only in the case of simple samples, which composition is stable and well known or when highly selective reagents are used. An isolation of an analyte from matrix is another solution for increasing the selectivity of assay. But every additional operation introduced into sample preparation procedure extents time and costs of single analysis and increases risk of loss or contamination of the analyte.

One of the simplest method for an increasing a selectivity is derivatisation of spectra. This operation allows to remove spectral interferences and as a consequence leads to increase selectivity of assay. Derivatisation of sets of digital data is well known method of separation useful signals from noised data [1]. Historically, the beginning of derivative spectrophotometry is dated on 1953 when the first analogue spectrophotometer was build by Singleton and Cooler [1]. But the fast development of this technique started in 70-s of twentieth century, when new generation of spectrophotometers controlled by computers were constructed. An apogee of its popularity occurred in 80-s of last century. Nowadays, it is only additional technique, rarely used, though it is fully available as a build-in function in software of modern spectrophotometers. I hope that this work gives some light on derivative spectrophotometry and restores it in some way.

2. Basic theory and properties of derivative spectrophotometry

Derivative spectrophotometry is a technique which is based on derivative spectra of a basic, zero-order spectrum. The results of derivatisation of function described a run of absorbance curve is called the derivative spectrum and can be expressed as:

$$^nD_{x,\lambda}=d^nA/d\lambda^n=f(\lambda) \quad \text{or} \quad ^nD_{x,v}=d^nA/dv^n=f(v)$$

where: n – derivative order, $^nD_{x,\lambda}$ or $^nD_{x,v}$ represents value of n-order derivative of an analyte (x) at analytical wavelength (λ) or at wavelength number (v), A- absorbance.

Derivative spectrophotometry keeps all features of classical spectrophotometry: Lambert-Beer law and law of additivity.

Lambert-Beer low in its differential form is expressed as:

$$^nD = \frac{d^nA}{d\lambda^n} = \frac{d^n\varepsilon}{d\lambda^n} \cdot c \cdot l$$

Where ε-molar absorption coefficient ($cm^{-1}mol^{-1}l$) , c – concentration of analyte (mol l^{-1}), l- thickness of solution layer (cm).

Derivative spectrum of **n**-component mixture is a sum of derivative spectra of individual components:

$$^nD_{mix}=^nD_1+^nD_2+...+^nD_n$$

A new feature of derivative spectrophotometry is a dependence of derivatisation results on geometrical characteristic of starting, zero-order spectrum. A shape and an intensity of the resulted derivative spectrum depend on half- heights width of peak in basic spectrum:

$$^nD=P^nA_{max}L^{-1}$$

where P^n- polynomial described run of n-derivative curve, n- derivative order, L- width of half- heights of peak of zero-order spectrum.

Due to this property broad zero-order spectra are quenched with generation of higher orders of derivatives while narrow undergo amplification. If the zero-order spectrum possess two bands A and B which differ from their half- heights width ($L_B>L_A$), after a generation of n-order derivative a ratio of derivatives intensity can be expressed as:

$$^nD_A/^nD_B=(L_B/L_A)^n$$

This dependence leads to increase in selectivity and/or sensitivity of assay. It allows to use for analytical properties a narrow band, overlapped or completely hooded by a broad ones.

The shape of derivative spectrum is more complicated than its parent one (Fig. 1). New maxima and minima appeared as results of derivatisation. The generation of **n-th** order derivative spectrum produces (n+1) new signals: an intense main signal and weaker bands, so called satellite or wings signals. Position of maxima or minima depend on order of derivative. The main extreme of derivative spectra of even order is situated at the same wavelength as maximum in zero-order spectrum, but for 2, 6 and 10-th order it becomes minimum in derivative spectrum and for 4, 8 and 12-th order it remains as a maximum (Fig. 1). The point of initial maximum converts into the point of inflection in derivative spectra of odd order. A narrowing of new signals is observed during generation of consecutive derivative spectra. This feature leads to narrowing bands and as a consequence to separation of overlapped peaks.

3. Generation of derivative spectra and their properties.

Modern software's controlled spectrophotometers allow not only acquisition and storage of registered spectra. They are equipped in modules enable mathematical operation like addition, subtraction, multiplication as well as derivatisation.

A registered UV-Vis absorption spectrum is a two-dimensional set of points with co-ordinates (λ, A), where λ – wavelength, A- absorbance. Derivative spectra can be obtained by direct calculation of ordinate increment or by fitting a function described a course of spectrum curve and next its derivatisation [1]. Another approach is to find a polynomial representing an absorption curve [1]. A proper form of the polynomial can be found if its all coefficients are known. If long set of n-data is disposed, determination of polynomial coefficients requires to solve **n** equations with **n**-unknowns. This is very hard, laborious job which could be impossible for long sets of data. There are many mathematical approaches simplifying this task [1]. The most popular is Savitzky-Golay algorithm [2] and its modifications [3,4]. Savitzky-Golay algorithm [2] does not analyze a whole set of points but only one exact point and its closest neighbourhood: **m** points from left and **m** points from right of the chosen neighbourhood of central point. A width of analysed set of points is equal **2m+1** and is called a derivatisation window. The coefficients of polynomial are calculated by the least square method for central point and next derivatisation window is moved right by one point and calculations for new central point are repeated. The result of this approach is a set of new points which creates a new – derivative spectrum. Usually the new set of points is shorter by **2m** points in comparison to the parent one. It isn't problem because the recorded spectrum usually is the long set of points and the clipped points are from beginning and end of zero order-spectrum which are useless from analytical point of view. Some improvements of Savitzky-Golay algorithm were done. Originally Savitzky-Golay algorithm was devoted for derivatisation of spectra with uniformly spaced sets of data. Nowadays it can be applied for nonuniformly spaced sequence [3]. There are modification which allow derivatisation without loss of extreme points [4].

The use of Savitzky-Golay algorithm requires optimalisation such parameters as derivative order, polynomial degree, width of derivatisation window and manner how derivative is generated. Analytical usefulness of resulted derivative spectrum depends on proper selection of mentioned parameters. Their selection should be done by taking into account a shape of initial zero-order spectrum and spectral properties of accompanied compounds.

- derivative order

Proper separation of overlapped signals can be achieved if appropriate derivative order is used. Optimal derivative order is a function of signals height, their width at half height and distance between maxima in basic spectrum [5,6]. It is recommended to use low orders if the basic spectrum is a sum of wide bands, while for the spectra consisted of narrow bands – higher orders. Generation of the high order derivative suppress very fast intensity of wide bands and magnify the narrow one[1, 5].

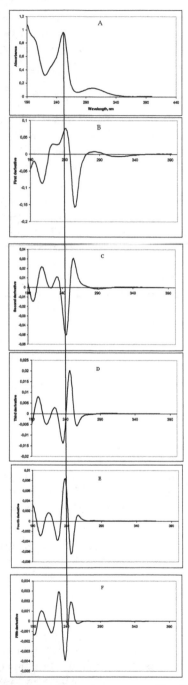

Fig. 1. Zero order (A) and consecutive derivative spectra (B→F) of aqueous solution of promazine hydrochloride (10 ppm); zero order spectrum has been recorded on Hewlett-

Packard HP-8452A diode array spectrophotometer with following working parameters: integration time 1 s, spectral bandwidth 2nm, spectrum scan 0.1 s. Derivative spectra were generated using Savitzky Golay algorithm by PC computer equipped with Excel for Microsoft Windows ($\Delta\lambda$=10 nm, second polynomial degree, derivatives of higher orders were generated by gradual derivatisation of derivative spectra of lower order).

- polynomial degree

The next optimized parameter is the polynomial degree. There is a similar dependence as in the case of derivative order. The high polynomial degrees should be used for spectral curves with sharp and narrow signals. Application of inappropriate polynomial degree gives a distorted derivative spectrum without useful analytical information [5]. In the case of multicomponent analysis, the use of polynomials of different degrees can allow to increase spectral differences of assayed compounds and their selective determination [5].

- width of derivatisation window

A proper selection of this parameter is crucial for quality and quantity of analytical information available in derivative spectrum. Application of the broad derivatisation window gives a smooth averaged derivative spectrum without spectral details. So, the broad derivatisation window is recommended for derivatisation of a zero-order spectra with broad irregular bands with a significant oscillatory constituent [5]. In the case of the basic spectrum with narrow absorption bands the narrow derivatisation window should be used. Otherwise the important analytical information could be lost and resulted maxima of derivative spectrum couldn't correspond to the real one[5].

- manner of generation of derivative spectra

Derivative spectra of higher orders can be obtained using Savitzky-Golay algorithm in two ways: by direct generation of desired derivative spectrum or by gradual generation of first order derivative on consecutive spectra: $^{0}A \xrightarrow{1} \dfrac{dA}{d\lambda} \xrightarrow{1} \dfrac{d(dA)}{d(d\lambda)} \xrightarrow{\dots} \dots \dfrac{d^{n}A}{d\lambda^{n}}$

It is very often observed that direct generation of the high-order derivative gives distorted analytically useless spectra (Fig. 2). Selection of derivatisation manner depends on shape of basic spectrum. It is recommended to apply the gradual derivatisation in the case of complicated zero-order spectra. A progressive generation of derivative spectra gives smooth derivative spectrum with advantageous signal-to-noise ratio.

Derivative spectrophotometry can be very useful additional tool which helps to solve some complicated analytical problems. Mathematical processing of spectra is very easy to use as modern spectrophotometers are computer controlled and their software are equipped in derivatisation unit. A proper selection of mathematical parameters gives profits in improved selectivity, sometimes sensitivity and in simplification of analytical procedure.

4. Analytical application of derivative spectrophotometry

Derivative spectrophotometry (DS) has found a wide application in quantitative chemical analysis. As the latest applications have been gathered and described in reviews published previously [8,9], this part is focused on the recent use of DS. Based on scientific literature the following fields of application of derivative spectrophotometry can be distinguished:

a. Multicomponent analysis. This group is the most numerous. The goal of proposed methods is application of DS for determination of one analyte in presence of matrix or for simultaneous assaying of few analytes.
b. Calculation of some physico-chemical constants, e.g. reaction, complexation or binding constants [10].
c. Application for investigation of some processes kinetics[11, 12].

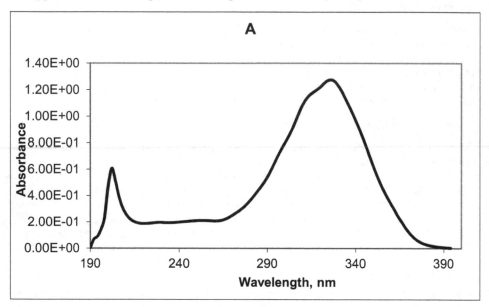

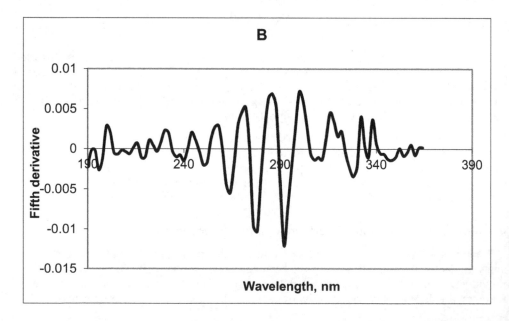

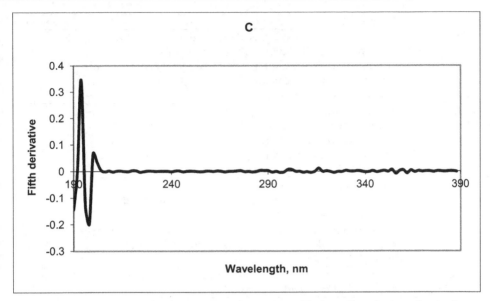

Fig. 2. Zero order (A) and fifth derivative order spectra of ethanolic solution of retinol acetate (10 ppm). Spectrum B has been obtained by gradual derivatisation, spectrum C by direct generation of fifth derivative from zero-order spectrum. Apparatus working conditions: Hewlett-Packard HP-8452A diode array spectrophotometer with following working parameters: integration time 1 s, spectral bandwidth 2nm, spectrum scan 0.1 s. Derivative spectra were generated using Savitzky Golay algorithm by PC computer equipped with Excel for Microsoft Windows ($\Delta\lambda$=14 nm, fourth polynomial degree).

a. Multicomponent analysis

Derivative spectrophotometry (DS) has been mainly used in pharmaceutical analysis for assaying of a main ingredient in a presence of others components or its degradation product. Pharmaceutical samples are characterised by high level of constituents and presence of a relatively simple and stable matrix. The spectral influences of disturbing compounds are easy to remove by derivatisation of spectra. The most numerous procedures based on derivative spectra have been devoted for determination of one components without sample purification. Another field of DS application is the use of it for simultaneous determination of two or more components. As a form of derivative spectrum is more complicated in comparison to its initial zero order, usually derivatives of low orders are employed for analytical purposes. Procedures used DS for pharmaceutical analysis are assembled in Tables 1 and 2.

b. Others applications

Derivative spectrophotometry was applied in different than pharmaceutical analysis areas of analysis. This method was utilised for the determination of amphothericin in various biological samples like plasma, serum, urine and brain tissue [40]. The combination of ratio spectra with their derivatisation allowed to remove spectral interferences caused by a presence of bilirubin in plasma [40].

Compound	Characteristic of the method	Reference
Sertraline	The proposed method is based on reaction of sertraline with chloranilic acid. First derivative spectrophotometry has been evaluated by measuring the derivative signal at 475.72 nm – 588.40 nm (peak to peak amplitude). Calibration graph was established for 5-100 µg mL^{-1} of sertraline	13
Estradiol valerate	The first-order derivative spectra were used for determination of estradiol in tablet. Measurements of derivative were made at 270 nm. The method showed specificity and linearity in the concentration range of 0.20 to 0.40 mg mL^{-1}.	14
Tropicamide	The measurements were carried out at wavelengths of 263.8 and 255.4 nm for third- and fourth-derivative, respectively. The method was found to be linear ($r^2 > 0.999$) in the range of 10-100 µg mL^{-1} for tropicamide in the presence of excipients. The method was applied for analyte determination in eye drops.	15
Nebivolol hydrochloride	Derivative spectrophotometry used for determination of nebivolol in bulk and in preparates.	16
Gemifloxacin mesylate	The proposed methods were based on the reaction of gemifloxacin with chloranilic acid and parachloranil to give highly coloured complexes. The coloured products were quantified spectrophotometrically at 530 nm and 540 nm at zero order, 590 and 610 nm for the first derivative and 630 and 650 nm for second order derivative. Beer's law was obeyed in the concentrations range of 10 to 60 µg mL^{-1}, 5 to25 µg mL^{-1} at zero order, 5 to 25 µg mL^{-1}, 5 to 40 µg mL^{-1} at first order and 2 to 20 µg mL^{-1} and 2 to 14 µg mL^{-1} at second order.	17
Olanzapine	The first derivative values measured at 222 nm and the second derivative values measured at 230 nm (n=6) were used for the quantitative determination of the drug . Calibration graphs were linear in the concentration range of olanzapine using 2-10 µg mL^{-1} for first and second derivative spectrophotometric method.	18
Galanthamine	The 1st derivative zero crossing spectrophotometry was proposed for determination of galanthamine. Absorbance was measured at 277.4 nm. It obeyed Lambert-Beer's law in the range of 30-80 µg mL^{-1}.	19
Doripenon	The first derivative spectrophotometry was used for determination of doripenem in pharmaceuticals in the presence of its degradation products. The Beer low was obeyed in the range (0.42-11.30)x10^{-2} mg L^{-1}.	20
Tropisetron	The first derivative spectra were applied for determination of analyte in the presence of its degradates. The quantification was done by measurement of first-derivative amplitude at 271.9 nm. The obtained results were in a good	21

Compound	Characteristic of the method	Reference
	agreement with those obtained by HPLC and TLC methods.	
Fluphenazine Pernazine Haloperidol Promazine	Derivative spectrophotometry was used for quantification of fluphenazine, pernazine and haloperidol in their preparations. First and second derivative were applied for determination of active ingredients in pharmaceutical preparations.	22
Ezetimibe	First, second and third derivative spectrophotometric methods were proposed and utilised for determination of ezetimibe in pharmaceuticals.	23
Oxybutynin hydrochloride	First derivative of ratio spectra was used for determination of analyte in presence of its degradation product.	24
Ertapenem	First derivative and first derivative of ratio spectra methods were applied for determination of ertapenem in the presence of its degradation product. The analyte was assayed by the first method at 316 nm in the range 4-60 µg mL^{-1}. The second method allowed ertapenem determination at 298 nm and 316 nm in the same concentration range using spectrum of degradant at 28 µg mL^{-1}as a divisor.	25

Table 1. Determination of one component in sample

Compound	Characteristic of the method	Reference
Democlocycline and minocycline	First derivative spectra were used for simultaneous determination of drugs in synthetic mixtures. The linearity in the ranges 10-40 µg mL^{-1} and 10-50 µg mL^{-1} were obeyed for democlocycline and minocycline, respectively. The method was applied for the analysis of these drugs in clinical samples, urine and honey.	26
Rupatadine and montelukast	The quantification was achieved using first-order derivative method. Rupatadine was determined at 273.46 nm, while montelukast at 297.27 nm. The method was applied for determination of both compounds in their combined dosage form.	27
Ambroxol and doxycycline	First derivative of ratio spectra method was applied for simultaneous determination of both analytes in pharmaceutical formulations and in laboratory-made mixtures.	28
Tramadol and ibuprofen	The first-derivative method was proposed for simultaneous determination of both compounds. The measurements of amplitude was done at 230.5 and 280 nm for tramadol (Trama) and ibuprofen (Ibu), respectively. The linearity was obeyed in the rage 5-50 µg mL^{-1} for Trama and 5-100 µg mL^{-1} for Ibu.	29
Sodium rabeprazole and itopride	First derivative of ratio spectra method was applied for simultaneous determination of both analyte. The amplitudes at 231 nm and 260 nm were used for	30

Compound	Characteristic of the method	Reference
hydrochloride	quantification of rabeprazole and itopride, respectively.	
Alprazolam and fluoxetine hydrochloride	Second derivative spectrophotometry (D2) was applied for simultaneous estimation of alprazolam (ALP) and fluoxetine hydrochloride (FXT) in pure powder and formulation. Quantitative determination of the drugs was performed at 232.14 nm and at 225.25 nm for ALP and FXT, respectively. Quantification was achieved over the concentration range 4-14 µg mL^{-1} for both drugs with mean recovery of 99.36 ± 0.84 and 99.60 ± 0.93 % for ALP and FXT, respectively.	31
Drotaverine hydrochloride and mefenamic acid	The second-order derivative spectra were used for simultaneous determination of drotaverine (DRO) and mefenamic acid (MEF). Calibration graphs were constructed over the concentration range of 4-24 µg/mL^{-1} for DRO and MEF. Detection and quantitation limit were 0.4348 and 1.3176 µg/mL^{-1} for DRO and 0.6141and 1.8611 µg/mL^{-1} for MEF. The method was applied for determination of both ingredients in combined dosage forms.	32
Triprolidine hydrochloride and pseudoephedrine hydrochloride	Second derivative spectrophotometric method was proposed for simultaneous determination of pseudoephedrine hydrochloride (PSE) and triprolidine hydrochloride (TRI). The second derivative amplitudes of PSE and TRI were measured at 271 and 321 nm, respectively. The calibration curves were linear in the range of 200 to 1,000 µg mL^{-1} for PSE and 10 to 50 µg mL^{-1}for TRI.	33
Clopidogrel bisulphate and aspirin	The method was based on the second-derivative spectra of both ingredients. The amplitude at 254.0 nm was used for clopidogrel bisulphate, while at 216.0 nm for aspirin. The linearity was obeyed in the range 5.0- 30.0 µg mL^{-1} for both compounds.	34
Simvastatin and ezetimibe	The first-order derivative spectrophotometric method was proposed for simultaneous determination of analytes in their mixtures. The measurements were carried out at 219 and 265 nm for simvastatin and ezetimibe respectively. The validation of method was done. The range of application was estimated to be 2-40 µg mL^{-1} for simvastatin in the presence of 10 µg mL^{-1} ezetimibe and 1-20 µg mL^{-1} of ezetimibe in the presence of 20 µg mL^{-1} of simvastatin.	35
Fe(III) and Al(III) ions	The proposed method was based on the first derivative spectra of Al^{3+} and Fe^{3+}complexes with chrome azurol S. The proposed procedure was successfully applied for simultaneous determination of studied ions in standard mixtures, pharmaceuticals and in post-haemodialysis samples.	36

Compound	Characteristic of the method	Reference
Calcium and magnesium ions	The reaction of studied ions with pyrogallol red at presence of Tween 80 was applied. Next the first and second derivative spectra of complexes were applied for quantification of calcium and magnesium in multivitamin preparations, samples of human serum and in drinking water.	37
Copper and palladium	The proposed method utilise the reaction of studied ions with morpholinedithiocarbamate (MDTC). Derivative spectra of generated complexes allowed simultaneous determination of Cu and Pd in pharmaceutical samples, synthetic mixtures, alloys and biological samples.	38
Paracetamol, propiphenazone and caffeine	First to fourth derivative spectra of components were subjected to chemometric analysis (principal component regression, PCR; partial least squares with one dependent variable, PLS-1; three dependent variables, PLS2) and adopted for multicomponent analysis. The third derivative spectra of all ingredients became a basis of quantification method.	39

Table 2. Application of DS for multicomponent analysis

The fourth-derivative spectra of molybdenum complexes of tetramethyldithiocarbamate (tiram) fungicide were used for its quantification in commercial samples and in wheat grains [41]. Atrazine and cyanazine were assayed in food samples by first- derivative spectrophotometry [42]. In order to improve results of assay, the first-derivative spectra of the binary mixture were subjected to chemometric treatment (classical least squares, CLS; principal component regression, PCR and partial least squares, PLS). A combination of first-derivative with PCR and PLS models were applied for determination of both herbicides in biological samples [42]. A first-derivative spectrophotometry was used as a reference method for simultaneous determination Brillant Blue, Sunset Yellow and Tartrazine in food [43].

First derivative of ratio spectra was applied for determination of strontium, magnesium and calcium in Portland cement [44]. The proposed procedure was based on complexation of studied ions with Alizarin Complexone.

As it is mentioned above, derivative spectrophotometry seems to be a very useful tool for physico-chemical studies. It can be applied for investigation of reaction kinetics [11,12], or for determination of chemical reaction constants.

First derivative spectra of levomepromazine (LV) and its sulphoxide were employed for investigation of LV photodegradation [11]. The degradation process of biapenem was monitored by measurement of first-derivative amplitude at 312 nm [12]. The determined rate constants for studied process were in good agreement with those obtained by HPLC method [12]. The second-order derivative spectrophotometric method was used for investigation of solvolytic reaction 2-phenoxypropionate ester of fluocinolone acetonide [45]. The run of process was observed by measurement the second-order amplitude at 274.96 nm corresponded to fluocinolone acetonide. The solvolysis rate constant was calculated using derivative method and compare with those obtained by HPLC methods [45].

An interesting application of derivative spectrophotometry was described by Wu and Zivanovic [46]. They proposed the use of the first derivative spectra for determination of the degree of acetylation of chitin and chitosan. They employed the evaluated procedure for commercial samples.

5. New trends in derivative spectrophotometry

The provided short review shows good and bad sides of derivative spectrophotometry. It has mainly found application in pharmaceutical analysis for control of pharmaceuticals. It gives good results for samples with well defined composition. A main compound usually is present in its commercial forms at a relatively high level, convenient for spectrophotometric determination. An application of derivative spectrophotometry simplified procedure and allows to determine an active compound in presence of matrix (others ingredients, its degradation products) without primary sample preparation.

An analysis of scientific articles shows new trends in the use of derivative spectrophotometry. First direction of development is a combination of derivative spectra with chemometric methods [28, 36, 39, 42]. Procedures based on derivatisation of ratio spectra [24, 25,28, 30, 40, 44, 47] belong to the same group. An interesting modification of derivative spectrophotometric procedure described Eskandari [48]. A fusion of H-point standard addition method with the first derivative of mixture spectra was applied for simultaneous determination palladium and cobalt. The method was applied for their determination in synthetic mixtures and alloys.

The second observed trend is an association of derivatisation with others instrumental methods. Every set of digital data can be subjected derivatisation. So this mathematical approach was applied for data processing with synchronous fluorescence spectroscopy. The second derivative synchronous fluorimetry was used for simultaneous determination of sulpiride and its degradation product [49]. For quantification were used amplitudes of ^2D peaks at 295.5 nm and at 342 nm corresponded to main compound and its degradate, respectively. The method was applied for studies of the kinetics of alkaline degradation of drug.

Kang et al. [50] developed the first derivative synchronous fluorescence method for simultaneous assay of traces of some polycyclic aromatic hydrocarbons in human urine. Proposed method was fast, sensitive, selective and reliable. The results were comparable with those obtained by HPLC method.

Derivative spectrophotometry was applied for resolving and quantification of overlapped peaks in capillary electrophoresis [51]. Derivatisation of electropherogram improved separation of compounds. An elaborated procedure was used for determination of eleven derivatives of benzoic acid.

6. Disadvantageous of derivative spectrophotometry

Specific properties of derivative spectrophotometry can be a source of an additional errors. As it is mentioned previously a shape of derivative spectrum is closely connected with the shape of its parent zero-order spectrum. Small changes in a course of curve describing basic spectrum are strongly magnified in derivative spectrum. Application of derivative spectrophotometry requires from analyst knowledge about its specific properties. The main disadvantage of derivative spectrophotometry is its poor reproducibility. It is result of strong dependence of derivative spectrum on recording parameters of used spectrophotometer like scan rate, spectral width of beam, integration time and interpoint distance[1, 5, 7]. Zero-order spectra of the same substance obtained on different spectrophotometers can be identical, but derivatisation of them gives different results. The generated derivative spectra can derived in intensity, shape and positions of maxima and minima. So restoration of given literature method requires to use the same type of apparatus with the same working parameters described in an article or reoptimisation parameters of method on an own spectrophotometer.

Optimisation of used working spectrophotometer parameters should be done when a new derivative-spectrophotometric method is elaborated. A construction of some spectrophotometers does not allow to check influence of whole factors, but if more advanced equipment is available it is worth to do.

As a result of derivatisation is closely connected with geometrical features of a zero-order spectrum, it is obvious that a method of spectrum registration is a key-point. The use of broad beams gives the averaged smoothed zero order spectra. Application of narrow beams results in intensification and narrowing of absorption bands. But from the other hands, the narrowing of monochromator' slit increases an effects connected with beams bending on edges of the slit. The edge phenomenon causes additional noises which are recorded with absorbance. So the absorption spectrum recorded with too narrow monochromator' slit can be distorted by high level of noise.

Interpoint distance of registered spectrum is very important parameter. Absorption spectrum obtained by spectrophotometer possess a digital structure which is the result of construction of a monochromator and a manner of registration. Spectra registered with large interpoint distance are averaged, flat without many spectral details.

A level of noise enclosed in zero order spectrum directly influences a quality of generated derivative spectrum. It was proved that spectra registered with low scan rates and long integration times are less biased by noise. This is advantageous if high order derivative are generated [7].

Taking into account above information, it is obvious that reproducibility of method based on derivative spectrophotometry depends on reproducibility of parameters of registration of zero-order spectra. So, adaptation of elaborated in another laboratory derivative spectrophotometric method, requires application the same working parameters as used by authors. But this problem is completely ignored by scientists. Based on analysis of articles concerned on application of derivative spectrophotometry it could be stated that working parameters of spectra registration are very rarely given [8]. There is noticeable lack of standardisation in description of procedures based on derivative spectra. Very often, authors of scientific articles give only information what model of apparatus they used without any details of its working parameters as well as algorithm for derivatisation of spectra. In this case the published procedure can be used only if our laboratory is equipped with the same model of spectrophotometer supplied with the same software. Otherwise verification of literature' method requires reoptimisation, adaptation to our conditions

A geometrical features of derivative spectra can be a source of analytical errors. A course of derivative curve is different than its initial spectrum. A main band gets narrowing but additional satellite bands appear. If basic spectrum of mixture is subjected to derivatisation a resulted derivative spectrum of mixture is a sum of derivative spectra of each individual components. New peaks in the final spectrum can be the result of addition or subtraction, so their intensity undergo amplification or reduction. Very often their positions are shifted in comparison to their position in derivative spectra of individual components. Some analytical information can be lost during derivatisation or new false peaks can be generated. A careful analysis of course of derivative spectra of components and their mixtures at different compositions should be done to avoid such errors. A selection of optimal derivatisation parameters should be make taking into account influence of others components on intensity of derivative peaks of determined analyte. This procedure seems to be time- and labour-consuming but gives good results. Properly done selection of mathematical parameters of derivatisation and instrumental parameters of spectral analysis allows to elaborate selective method of determination and leads to minimise errors connected with features of derivative spectra.

7. Conclusions

Nowadays, derivative spectrophotometry is fully available with software's controlling modern spectrophotometers. Analysts receive an elegant tool which allows extraction of analytically useful information from spectra. An understanding of specific features of this technique and its proper utilisation leads to simplification of procedure and to increase a selectivity of assay.

8. References

[1] G. Talsky, Derivative Spectrophotometry, 1st ed., VCH, Weinheim, 1994.
[2] A. Savitzky, M. J. E.Golay, Anal. Chem. 36, 1627-1642 (1964).
[3] P. A. Gorry, Anal. Chem. 63, 534-536 (1991).
[4] P. A. Gorry, Anal. Chem. 62, 570-573 (1990).
[5] S. Kuś, Z. Marczenko, N. Obarski, Chem. Anal. (Warsaw) 41, 899-929 (1996).
[6] T. C. O'Haver, G. L. Green, Anal. Chem. 48, 312-318 (1976).
[7] T. C. O'Haver, T. Begely, Anal. Chem. 53, 1876-1878 (1981).
[8] J. Karpinska, Talanta 64, 801-822 (2004).
[9] F. Sanchez Rojas, C. Bosch Ojeda, Anal. Chim. Acta 635, 22-44 (2009).
[10] M. Gumustas, S. Sanli, N. Sanli, S.A.Ozkan, Talanta 82, 1528-1537 (2010).
[11] J. Karpinska, A.Sokol, M. Skoczylas, Spectrochim. Acta Part A 71, 1562-1564 (2008).
[12] J. Cielecka-Piontek, A. Lunzer, A. Jelinska, Cent. Eur. J. Chem. 9, 35-40 (2011).
[13] Y.F. M. Alqahtani, A.A. Alwarthan, S. A. Altamrah, Jordan J. Chem. 4, 399-409 (2009).
[14] A.S. L. Mendez, L.Deconto, C. V. Garcia, Quim. Nova, 33, 981-983 (2010).
[15] E. Souri, M. Amanlou, S. Shahbazi, M. Bayat, Iranian J. Pharm. Sci. 6, 171-178 (2010).
[16] S. M. Malipatil, P.M. Deepthi, S.K.K. Jahan, Int. J. Pharm. Pharmaceut. Sci. 3, 975-1491, (2011).
[17] D. Madhuri , K.B. Chandrasekhar, N. Devanna, G. Somasekhar, Int. J. Pharm. Sci. Res. 1, 222-231 (2010).

[18] V.M. Patel, J. A. Patel, S.S. Havele, S.R. Dhaneshwar, Int. J. Chem. Tech. Res. 2, 756-761, (2010).

[19] K. Mittal, R. Kaushal, R. Mashru, A. Thakkar, J. Biomed. Sci. Eng. 3, 439-441 (2010).

[20] J. Cielecka-Piontek, A. Jelińska, Spectrochim. Acta Part A. 77, (554-557 (2010).

[21] S. L. Abdel-Fattah, A. Z. El-Sherif, K. M. Kilani, D. A. El-Haddad, J. AOAC Int. 93, 1180-1191 (2010).

[22] M. Stolarczyk, A. Apola, J. Krzek, A. Sajdak, Acta Pol. Pharm. 66, 351-356 (2009).

[23] M. Sharma, D.V. Mhaske, M. Mahadik, S.S. Kadam, S. R. Dhaneshwar, Indian J. Pharm. Sci. 70, 258-260 (2008).

[24] N. E. Wagiem, M.A. Pegazy, M. Abdelkawy, E. A. Abdelaleem, Talanta 80, 2007-2015 (2010).

[25] N.Y.Hassan, E. M. Abel-Moety, N.A. Elragey, M.R. Rezk, Spectrochim. Acta Part A 72, 915-921 (2009).

[26] A.R.G.Prasad, V. S. Rao, Sci. World 8, 34-38 (2010).

[27] P.Patel, V. Vaghela, S. Rathi, N. Rajgor, V. Bhaskar, J. Young Pharm. 1, 354-360 (2009).

[28] G. M. Hadad, A. El-Gindy, W. M. M. Mahmoud, Spectrochim. Acta Part A 70, 655-663 (2008).

[29] A. B. Thomas, N. G. Dumbre, R. K. Nanda, L. P. Kothapalli, A.A. Chaudari, A.D. Deshpande, Chromatographia 68, 843-847 (2008).

[30] S. S. Sabnis, N. D. Dhavale, V. Y. Jadhav, S. V. Gandhi, Spectrochim. Acta Part A 69, 849-852 (2008).

[31] R. B. Patel, M. R. Patelb, M. B. Shankara, K. K. Bhattb, Eurasian J. Anal. Chem. 4,76-86 (2009).

[32] J. D. Patel, B. A. Patel, B. P. Raval, V. M. Vaghela, J. Pharm. Res, 3,566-569, (2010).

[33] L. Sriphong, A. Chaidedgumjorn, K. Chaisuroj, World Acad. Sci., Eng. Technol. 55, 573-577 (2009).

[34] M. D.Game, D. M. Sakarkar , Inter. J. Chem. Tech. Res. 2, 1886-1891 (2010).

[35] E. Souri, M. Amanlou, E-Journal of Chemistry, http://www.e-journals.net, 7(S1), S197-S202 (2010).

[36] N. Aguerssif, M. Benamor, M. Kachbi, M.T. Draa, J. Trace Elements Med. Biol. 22, 175-182 (2008).

[37] M. Benamor, N. Aguerssif, Spectrochim. Acta Part A 69, 676-681 (2008).

[38] V. Kaur, A. K. Malik, N. Verma, Anal. Letters 40, 2360-2373 (2007).

[39] M. De Luca, F. Oliviero, G. Ioele, G. Ragno, Chemometr. Intell. Lab. Sys. 96, 14-21 (2009).

[40] J.S. Millership, F.McCaffrey, D. Tierney, J. Pharm. Biomed. Anal. 48, 408-413 (2008).

[41] V. K. Sharma, J. S. Aulakh, A.K. Malik, Talanta 65, 375-379 (2005).

[42] G. Zhang, J. Pan, Spectrochim. Acta Part A 78, 238-242 (2011).

[43] T.M.Coelho, E. C. Vidotti, M. C. Rollemberg, A. N. Medina, M. L. Baesso, N. Celle, A. C. Bento, Talanta 81, 202-207 (2010).

[44] K. A. Idriss, H. Sedaira, S.S. Ahmed, Talanta 78, 81-87 (2009).

[45] B. Markovic, S. Vladimirov, O. Caudina, V. Savic, K. Karljikovic-Rajic, Spectrochim. Acta Part A 75, 930-935 (2010).

[46] T. Wu, S. Zivanovic, Carbohyd. Polym. 73, 248-253 (2008).

[47] F. A. El-Yazbi, H. H. Hammud, S. A. Assi, Spectrochim. Acta Part A 68, 275-278 (2007).

[48] H. Eskandari, Spectrochim. Acta Part A 63, 391-397 (2006).

[49] A. Abdeal, N. El-Enany, F. Belal, Talanta 80, 880-888 (2009).

[50] R-H. Kang, Y-S. Wang, H-M. Yang, G-R. Li, X. Tan, J-H. Xue, J-Q. Anal. Chim. Acta 658, 180-186 (2010).

[51] J. Zhang, Y. Hu, J. Liu, Z.Hu, Microchim. Acta 164, 487-491 (2009).

A Comparative Study of Analytical Methods for Determination of Polyphenols in Wine by HPLC/UV-Vis, Spectrophotometry and Chemiluminometry

Vesna Weingerl[*]

University of Maribor, Faculty of Agriculture and Life Sciences, Hoče, Slovenia

1. Introduction

Wine, especially red wine, is a very rich source of polyphenols, such as flavanols (catechin, epicatechin, etc.), flavonols (quercetin, rutin, myricetin, etc.), anthocyanins (the most abundant is malvidin-3-o-glucoside), oligomeric and polymeric proanthocyanidins, phenolic acids (gallic acid, caffeic acid, p-coumaric acid, etc.), stilbenes (*trans*-resveratrol) and many others polyphenols. Many of these compounds (e.g. resveratrol, quercetin, rutin, catechin and their oligomers and polymers proanthocyanidins) have been reported to have multiple biological activities, including cardioprotective, anti-inflammatory, anti-carcinogenic, antiviral and antibacterial properties (King et al., 2006; Santos- Buelega & Scalbert, 2000). These biological properties are attributed mainly to their powerful antioxidant and antiradical activity.

Regular, moderate consumption of red wine reduced the incidence of many diseases such as risk of coronary heart disease (CHD), atherosclerosis, cancers, etc. (Cooper et al., 2004; Opie & Lecour, 2007). The most intriguing are the studies which reported the possible association between red wine consumption and decrease in risk, and some suppression and inhibition of cancers (Briviba et al., 2002). Currently, chemoprevention is being used in medicine as a new strategy to prevent cancers. Natural phytochemicals, including red wine polyphenols, appear to be very promising substances to block, reverse, retard or prevent the process of carcinogenesis (Russo, 2007). Many epidemiological studies have found that regular intake of red wine or red wine polyphenols has positive effects on human health. Therefore, determination of the chemical composition, polyphenols content and antioxidant activity of red wine could be very useful for the interpretation of epidemiological studies.

Phenolic antioxidants define total antioxidant potential of wines and have the greatest influence on it. Authors showed that in grape seeds gallic acid, catechins and epicatechins prevailed, whereas in peel ellagic acid, quercetin and trans-resveratrol were most common.

---

[*] Corresponding Author

The high antioxidant potential of red wines can be ascribed to the synergistic effect of the mixture of natural phenolic antioxidants (Lopez-Velez et al., 2003).

Production technology is one of the main factors influencing the high antioxidant potential of red wines (Downey et al., 2006; Vršič et al., 2009). During winemaking the grape pulp is fermented, and fruit peel and seeds are very rich in phenolic antioxidants. The concentration of polyphenols in peel is higher than in the flesh (Darias-Martin et al., 2000; Fuhrman et al., 2001). During intensive pressing or during long contact of juice with pulp the content of phenolic compounds increases rapidly (Fuhrman et al., 2001). It was found that in wines fermented with peel the concentration of phenolic antioxidants was 2 times higher than in wines fermented without peel (Darias Martin et al., 2000).

Antioxidant potential and polyphenol composition were assessed in wine of Croatian origin (Katalinic et al., 2004). The concentration of total polyphenols in red wines ranged from 2200 to 3200 mg gallic acid per litter (mg GA/L).

In winemaking, phenolic antioxidants are extracted from berry skins, seeds and stems during crushing and fermentation. Due to the market demand, knowledge of the concentration of phenolic antioxidants in wine and their antioxidant potential is very important.

Wine, especially red wine, is a very rich source of flavonol quercetin and many others polyphenols. Various methods for characterisation of total antioxidant potential are presently in routine use, although some are non-stoichiometric (Alimelli et al., 2007; Campanella et al., 2004; Careri et al., 2003; Carralero Sanz et al., 2005; De Beer et al., 2005; Fernandez-Pachon et al., 2004; Giovanelli, 2005; Gomez-Alonso et al., 2007; Magalhaes et al., 2009; Makris et al., 2003; Malovana et al, 2001; Mozetič et al., 2006; Prior et al., 2005; Prosen et al., 2007; Recamales et al., 2006; Spigno & De Faveri, 2007; Staško et al., 2008; Weingerl et al., 2009; Woraratphoka, 2007). Some of these methods allow for rapid characterization of wines, and allow for evaluation of synergistic effects of various wine components, e.g. transition metals (Strlič et al., 2002), which can have pro-oxidative effects in a mixture with phenolic compounds. The results of such analyses are usually given in equivalents of gallic acid or other reference compounds. Among these methods, determination of total phenolic content using the Folin-Ciocalteu reagent, as described by Singleton and Rossi (Singelton & Rossi, 1965), is very common.

Considering the accumulated knowledge on the effect of phenolic antioxidants on human health and the resulting market requirements it is highly important to have well developed, robust and established methods for their determination (Minussi et al., 2003; Urbano-Cuadrado et al., 2004).

In this study we compared three analytical methods: high pressure liquid chromatography (HPLC) with UV-vis detection, UV-vis spectrophotometry and chemiluminometry.

For separation and determination of phenolic acids and flavonoids, HPLC is the established technique (Nave et al., 2007; Rodriguez-Delgado et al., 2001; Spranger et al., 2004; Vitrac et al., 2002). The chromatographic conditions include the use of, almost exclusively, a reversed phase C18 column; UV-vis diode array detector, and a binary solvent system containing acidified water and a polar organic solvent (Tsao & Deng, 2004).

A Comparative Study of Analytical Methods for Determination of Polyphenols in Wine by HPLC/UV-Vis,
Spectrophotometry and Chemiluminometry

69

With use of HPLC we can reach separation of non-stabile and heavy volatile analytes on the base of different chemical interactions of the analytes with mobile phase and stationary phase. We use a non-polar stationary and polar mobile phase (reversed phase chromatography). HPLC with UV-visible detection was used for determination of antioxidant compounds content of gallic acid, (+)-catechin, (-)-epicatechin, *trans*-resveratrol, *cis*-resveratrol and quercetin in numerous wine samples. The selected phenolic compounds are the most important wine antioxidants. Gallic acid, the main hydroxybenzoic acid in red wines, is a very potent antioxidant with three free hydroxyl groups. Because of the relatively slow extraction of gallic acid from grape seeds, higher concentrations are obtained with longer maceration times, which is characteristic for red wines. As the most important flavonol, quercetin was also included in our research.

Spectrophotometric determination of total antioxidant potential (TAP_{SP}) was performed with oxidation of phenolic compounds with Folin-Ciocalteu reagent after spectrophotometric method, described by Singelton in Rossi. Gallic acid was used as modelling solution.

Chemiluminometric determination of polyphenols was another possibility for evaluation of the total antioxidant potential in wine. Chemiluminescence, the emission of light as a consequence of relaxation of kind, which it is evoked between chemical reactions, has become very useful technique for studying oxidation of organic materials (Costin et al., 2003; Garcia-Campana & Baeyens, 2001; Hötzer et al., 2005; Kočar et al., 2008; Kuse et al., 2008). ABEL® (analysis by emitted light) antioxidant test kit, which contains photo protein Pholasin®, was used. Photo protein Pholasin® is the protein-bound luciferin from the bivalve mollusc Pholas dactylus, which reacts with luciferase and molecular oxygen to produce light (Knight, 1997; Michelson, 1978; Roberts et al., 1987). In reaction system substrate-catalyser-oxidant we can therefore inhibit occurrence of chemiluminescence with antioxidants. If there are antioxidants in the sample capable of scavenging superoxide, then these antioxidants will compete with Pholasin® for the superoxide and less light will be detected. Control samples containing no antioxidants were running with each assay. With measuring of decrease of intensity of chemiluminescence we evaluate total antioxidant potential (TAP_{CL}) of the sample (Hipler & Knight, 2001)..

The reactions of Pholasin® have been studied extensively (Dunstan et al., 2000; Müller et al., 1989; Reichl et al., 2000). It was found to have a 50- to 100-fold greater sensitivity towards superoxide than luminol. In addition, the decay of the Pholasin® chemiluminescent product was more rapid than that of the luminol product, leading to a greater accuracy in real-time kinetic studies. For these reasons, Pholasin® offers several advantages over luminol (Roberts et al., 1987). The luminescence of Pholasin® elicited with luciferase has a maximum at 490 nm, and that with Fe^{2+} shows a maximum at 484 nm (Shimomura, 2006).

In addition, we compared the results of the chromatographic method with those obtained using the spectrophotometric and chemiluminometric method.

In order to evaluate their potential individual contributions to TAP, phenolic antioxidants were analysed as pure solutions in the same concentration. The order of contributions of individual phenolic antioxidants to TAP determined according to spectrophotometric method was different than those determined with chemiluminometric method. Generalised, *cis*-resveratrol has the biggest contribution to TAP, following by *trans*-resveratrol, (-)-

epicatechin, (+)-catechin and quercetin. Interesting, gallic acid, as modelling solution by spectrophotometric determination of TAP_{SP}, shows the lowest contribution to TAP.

2. Experimental

137 wine samples (73 red, 54 white and 10 rosé wines) were purchased from wineries and directly analysed. Most of the red wine samples were from grape varieties Blue Frankish, Merlot, Cabernet Sauvignon, Pinot Noir, Refošk and Barbera. Cviček was used in the study as a typical Slovenian mixture of red and white wines, whereas from among the white wines, most samples were from grape varieties Welsh Riesling, Chardonnay, Traminer, Rhine Riesling, Yellow Muscat and Sauvignon. Vintages ranged from 1997 to 2006.

2.1 HPLC- UV/VIS method

The HPLC system Waters 600E was composed of isocratic pump W600, autosampler and Waters 996 photodiode array detector. The HPLC column Synergi Hydro RP 150 × 4.6 mm, 4 μm (Phenomenex, Torrance, California, USA) was used at 35 °C. Wavelengths of detection: (+)-catechin and (-)-epicatechin 210 nm, quercetin 253 nm, gallic acid 278 nm and *trans*-resveratrol 303 nm. Gallic acid, (+)-catechin hydrate, (-)-epicatechin, *trans*-resveratrol and quercetin dihydrate were purchased from Sigma-Aldrich (St. Louis, USA). All reagents and standards were prepared using Milli Q deionized water (Millipore, Bedford, USA).

The experimental conditions were: mobile phase A: 0.1% H_3PO_4; mobile phase B: MeOH; gradient elution: 0 min 90% A, 10% B; 15 min 78% A, 22% B; 25 min 50% A, 50% B; 34 min 34% A, 66% B; 35 min 90% A, 10% B for reconditioning of the system (8min); flow rate: 1.0 mL/min; injection volume: 50 μL; MeOH and ortophosphoric acid were HPLC-grade (Fluka, St. Gallen, Suisse) and were filtered and degassed before their use. The wine samples were diluted ten times with the respective mobile phases described above.

Stock solutions of standards were diluted in the mobile phase to obtain working standard solutions. Concentrations of the analytes were calculated from chromatogram peak areas on the basis of calibration curves. The method linearity was assessed by means of linear regression of the mass of analyte injected vs. its peak area. The repeatability was expressed as standard deviation (SD) of three separate determinations.

Typical standard deviations for determinations of the sum of phenolic antioxidants determined using HPLC/UV-vis were 0.015 mmol/L for red wines, 0.004 mmol/L for rosé and 0.050 mmol/L for white wines.

2.2 Spectrophotometry

Spectrophotometric determination of total antioxidant potential (TAP_{SP}) was performed according to the Singleton-Rossi procedure (Singelton & Rossi, 1965). TAP_{SP} of an antioxidant sample was estimated by measuring its reducing capacity with the Folin-Ciocalteu reagent using a spectrophotometer. The Folin-Ciocalteu reagent is a mixture of phosphowolframic acid ($H_3PW_{12}O_{40}$) and phosphomolybdenic acid ($H_3PMo_{12}O_{40}$), the absorbance of which was measured after the reaction at 765 nm using a Cary 1E spectrophotometer (Varian, California, USA). The Folin–Ciocalteu reagent was purchased from Merck (Darmstadt, Germany). It contains sodium tungstate, sodium molybdate,

A Comparative Study of Analytical Methods for Determination of Polyphenols in Wine by HPLC/UV-Vis, Spectrophotometry and Chemiluminometry

71

ortophosphoric acid, hydrochloric acid, lithium sulphate, bromine, hydrogen peroxide (Folin & Ciocalteu, 1927).

Briefly, 25 μL of a red and rosé wine sample or 250 μL of a white wine sample, 15 mL of distilled water, 1.25 mL of the diluted (1:2) Folin–Ciocalteu reagent, 3.75 mL of a sodium carbonate solution (20%) were mixed and distilled water was added to make up the total volume of 25 mL. The solution was agitated and left to stand for 120 min at room temperature for the reaction to take place. The calibration curve was prepared with gallic acid solutions in the concentration range from 0 to 1000 mg/L. The results are expressed as mmol gallic acid per litter (gallic acid equivalents - GAE).

The results for standards were highly reproducible (calibration curve squared regression coefficient >0.9993). All determinations were performed in triplicates. Typical standard deviation for spectrophotometric determinations of total phenolic content (TAP$_{SP}$) was 0.10 mmol/L for red wines, 0.09 mmol/L for rosé and 0.02 mmol/L for white wines.

2.3 Chemiluminometry

The Abel®-21 M2 antioxidant test kit (Knight Scientific Limited, Plymouth, UK) was used for chemiluminometric determination of total antioxidant potential (TAP$_{CL}$). Superoxide, generated in a tube containing Pholasin® with and without a sample of wine with unknown antioxidant potential leads to appearance of chemiluminescence, which was measured using a micro plate luminometer model Lucy (Anthos Labtec Instruments, Wals, Austria). Pholasin® is a bioluminescent photo protein of *Pholas dactylus*, which is a marine, rock-boring, bivalve mollusc. Antioxidant test kit used contains assay buffer (pH 7.2).

The Antioxidant Test procedure for Superoxide is provided by the supplier of the test kit (Hipler and Knight, 2001). The amount of sample was optimised to obtain not more than 90% and typically 50% signal inhibition. This signal was then corrected for sample dilution: 10 μL of sample was used, however, red wines and rosé wines were first diluted with water (1:10) while white wines were not.

The results are calculated as TAP$_{CL}$, expressed as % signal inhibition. Typical measurement uncertainty was 0.024 mmol/L for red wines, 0.016 mmol/L for rosé and 0.002 mmol/L for white wines.

2.4 Statistical analysis

Measurements are expressed as means ± standard deviations (SD) for three replicate determinations. Multivariate analysis was performed using SPSS 17.0 for Windows (SPSS Inc., Chicago, USA). As a typical data reduction and visualisation technique, principal component analysis (PCA) was used.

3. Results and discussion

Comparison of determinations of total antioxidant potential in different wines was performed with spectrophotometric and chemiluminometric method, while comparison of antioxidant compounds content in the same samples was performed using HPLC with UV/VIS detection.

Sum of phenolic compounds, determined using HPLC with UV-vis detection, summarized six manly phenolic compounds (gallic acid, (+)-catechin, (-)-epicatechin, *trans*-resveratrol, *cis*-resveratrol and quercetin). Regarding the knowledge about the influence of phenolic antioxidants on human health, selected phenolic compounds are the most important antioxidants of red wines. Gallic acid, the main hydroxybenzoic acid in red wines, is with three free hydroxyl groups a very strong antioxidant. Because of relative slow extraction of gallic acid from grape seeds, we obtain higher concentrations with longer maceration time, which is characteristic for red wines. In the group of non-flavonoid phenols we analysed also the main representative of stilbenes – resveratrol. Resveratrol (3,5,4'-trihydroxystilbene) is present in wine in four forms; we ware determining *trans*-resveratrol and *cis*-resveratrol. In the group of flavonoid phenols we choose isomers (+)-catechin and (-)-epicatechin, more specific, flavan-3-ols. As most important flavonol we include in our research quercetin. Chemically are flavonols 3-glycosides.

Relative good correlation is result between sum of phenolic compounds determined with HPLC/UV-vis and spectrophotometric determinations of total antioxidant potential TAP_{SP} ($r^2 = 0.91$) (Weingerl et al., 2009).

Light emitted by chemiluminescent substrate was expressed in relative light units (RLU). The time dependence of light intensity was measured and the peak intensity was converted into percent of inhibition relative to control. We expressed the results as total antioxidant potential, expressed as percent inhibition: $TAP_{CL} = ((max\ RLU_{control}) - (max\ RLU_{sample}))\ 100\ /\ (max\ RLU_{control})$. Figure 1 shows the correlation between sum of phenolic compounds

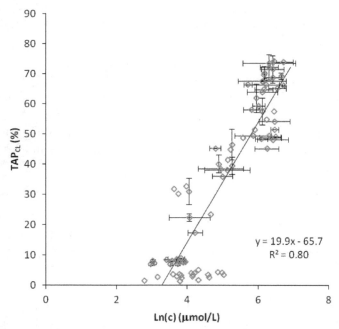

Fig. 1. Comparison between natural logarithm of sum of gallic acid, (+)-catechin, (-)-epicatechin, *trans*-resveratrol, *cis*-resveratrol and quercetin, determined using HPLC and TAP_{CL}. All wine samples included.

determined with HPLC/UV-vis and total antioxidant potential TAP_{CL} determined with chemiluminometry.

Sum of phenolic compounds, determined with HPLC method, correlate well with total antioxidant potential TAP_{CL}, determined with chemiluminometric method (r^2 = 0.80) (Fig 1).

With use of HPLC method we have the information about single phenolic antioxidants content, which doesn't consider synergistic influences between phenolic compounds in wine. Those effects considered only methods for determining total antioxidant potential. On the other hand, matrix effects may lead to different results obtained using other methods. In order to obtain a better insight into the extent of these effects, we compared all used methods.

To estimate the quantification limits, weighted tolerance intervals were used. The limit of quantification (in response units) is defined as 10 times the standard deviation at the lowest detectable signal (LC) plus the weighted intercept. The corresponding concentration LQ can be obtained by $LQ = (yQ - a)/b$, where a is the intercept, b is the slope of correlation curve (Zorn et al. 1997). The quantification limit for TAP_{SP} was calculated from data represented in Fig. 2, and amounts to 844 µmol/L gallic acid (Weingerl et al., 2011).

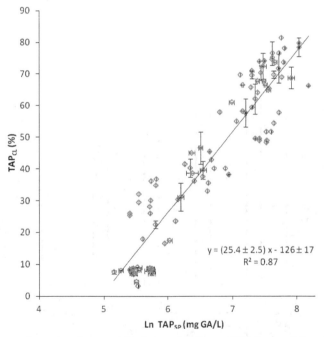

Fig. 2. Comparison of spectrophotometric determinations of TAP_{SP} and chemiluminometric results (TAP_{CL}).

For comparison, Table 1 shows further limits of quantification for determination of sum of HPLC determined phenolic antioxidants, TAP_{CL} and TAP_{SP} in wine.

Method	LOQ	Unit
Chemiluminometric method	32	µmol/L ascorbic acid
Sum HPLC/UV-vis	27	µmol/L
Spectrophotometric method	844	µmol/L gallic acid

Table 1. Limits of quantification (LOQ) for determination of polyphenols in wine.

As it is evident from Table 1, the chemiluminometric method exhibits similar quantification limits to the HPLC method for the determination of the sum of gallic acid, (+)-catechin, (-)-epicatechin, *trans*-resveratrol, *cis*-resveratrol and quercetin in wines (32 µmol/L). This is much lower quantification limit comparable with the spectrophotometric method, for the determination of total polyphenols, expressed in µmol/L of gallic acid.

Like total antioxidant potential determined with spectrophotometric or chemiluminometric method, like sum of six determined phenolic compounds are higher by red wines than white wines, rosé wines are giving intermediate results.

Evaluation of individual contributions of selected phenolic antioxidants to total antioxidant potential shows different results for spectrophotometric and chemiluminometric method (Fig.3 and Fig.4).

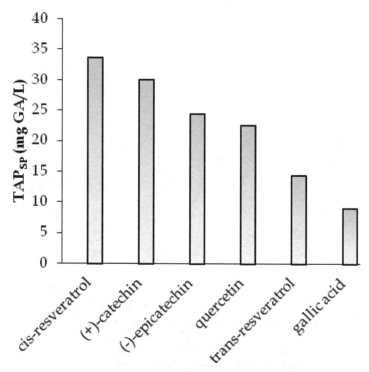

Fig. 3. TAP_{SP} of individual phenolic antioxidant solutions (50 µmol/L).

A Comparative Study of Analytical Methods for Determination of Polyphenols in Wine by HPLC/UV-Vis, Spectrophotometry and Chemiluminometry

75

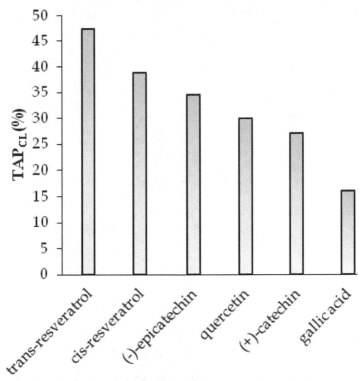

Fig. 4. TAP_{CL} of individual phenolic antioxidant solutions (50 μmol/L).

Regarding their antioxidant potential, *cis*-resveratrol has been found to be the most powerful scavenger among the analysed phenolic antioxidants. *cis*-resveratrol, (+)-catechin, (-)-epicatechin, quercetin, *trans*-resveratrol and gallic acid was the order of contributions to TAP_{SP}, determined according the spectrophotometric method. *trans*-resveratrol, *cis*-resveratrol, (-)-epicatechin, quercetin, (+)-catechin and gallic acid was the order of contributions to TAP_{CL}, determined according to chemiluminometric method. Gallic acid has, interesting, the lowest contribution to TAP. Generalised, *cis*-resveratrol has the biggest contribution to TAP, following by *trans*-resveratrol, (-)-epicatechin, (+)-catechin and quercetin.

To reduce the number of variables and to investigate the extent of correlation between the six individual phenolic antioxidants and total antioxidant potentials, determined with spectrophotometric and chemiluminometric method, principal component analysis (PCA) was performed. While 57.5% of the variation is explained by PC1 and another 11 % by PC2, we compared the loading factors in Figure 4 to investigate how the different variables might be co-correlated. TAP_{SP} and amount of gallic acid, determined with HPLC/UV-vis method are very strongly co-correlated and most strongly affected by quercetin content.

As is evident from Figure 4, there is a very strong co-correlation between TAP_{CL} and content of *cis*-resveratrol. Referred to this co-correlation TAP_{CL} is also strongly correlated to further antioxidants present especially in red grape varieties: (+)-catechin, (-)-epicatechin and *trans*-resveratrol.

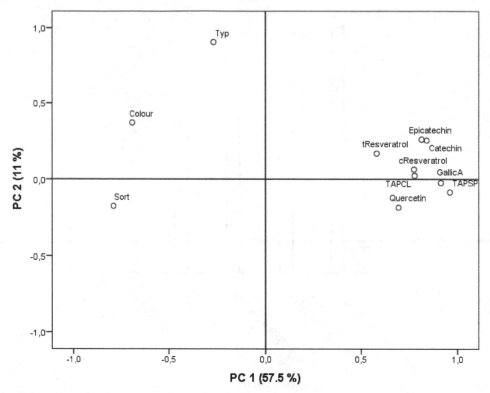

Fig. 5. Loading plot for PCA performed with all measured variables: single phenolic antioxidants, TAP$_{SP}$ and TAP$_{CL}$.

4. Conclusions

In this work, we compared two routinely used methods of determination of the following wine phenolic antioxidants: gallic acid, (+)-catechin, (-)-epicatechin, *trans*-resveratrol, cis-resveratrol and quercetin. We used liquid chromatographic method with UV-vis detection, the conventional Folin-Ciocalteu spectrophotometric method and chemiluminometric method. The comparisons led us to the following conclusions:

-   Determinations of the total antioxidant potential determined with chemiluminometric method correlate well with the sum of individual contents of gallic acid, (+)-catechin, (-)-epicatechin, *trans*-resveratrol, *cis*-resveratrol and quercetin as obtained using HPLC method in all wine samples.
-   *cis*-resveratrol has been found to be the most powerful scavenger among the analysed phenolic antioxidants.

Information about single phenolic antioxidants content doesn't consider synergistic influences between phenolic compounds in wine, we ware limiting only on components, which ware analysed individually. Information about content of antioxidant components also doesn't consider prooxidative influence of transition metals content. Those effects considered both described methods for determining total antioxidant potential (TAP$_{SP}$ and TAP$_{CL}$), on which indicates quality of correlation determining according to these two methods.

- PCA was applied to investigate the correlations between TAP$_{SP}$, TAP$_{CL}$ and individual phenolic antioxidants and especially good co-correlation was found with TAP$_{CL}$ and *cis*-resveratrol. TAP$_{SP}$ and amount of gallic acid, determined with HPLC method are very strongly co-correlated and most strongly affected by quercetin content.

If we consider content of individually determined phenolic antioxidants and not only antioxidative capacity of wine, we may use only HPLC method with UV-vis detection.

5. Acknowledgements

I offer my regards and blessings to all of those precious people who supported me in any respect during my academic growth.

6. References

Alimelli, A., D. Filippini, R. Paolesse, S. Moretti, G. Ciolfi, A. D'Amico, I. Lundström, C. Di Natale. (2007). Direct quantitative evaluation of complex substances using computer screen photo-assisted technology: the case of red wine. *Anal. Chim. Acta*, 597, 103-112.

Briviba, K., L. Pan and G. Rechkemmer. (2002). Red wine polyphenols inhibit the growth of colon carcinoma cells and modulate the activation pattern of mitogen-activated protein kinases. *The Journal of Nutrition.* 132, 2814-2818.

Campanella, L., A. Bonanni, E. Finotti, M. Tomassetti. (2004). Biosensors for determination of total antioxidant capacity of phytotherapeutic integrators: comparison with other spectrophotometric, fluorimetric and voltammetric methods. *Biosens. Bioelectr.* 19, 641-651.

Careri, M., C. Corradini, L. Elviri, I. Nicoletti, and I. Zagnoni. (2003). Direct HPLC Analysis of Quercetin and trans-Resveratrol in Red Wine, Grape, and Winemaking Byproducts, *J. Agric. Food Chem.* 51, (18), 5226-5231.

Carralero Sanz, V., M. Luz Mena, A. González-Cortés, P. Yáñez-Sedeño, J. M. Pingarrón. (2005). Development of a tyrosinase biosensor based on gold nanoparticles-modified glassy carbon electrodes. Application to the measurement of a bioelectrochemical polyphenols index in wines, *Anal. Chim. Acta*, 528, 1-8.

Cooper, K. A., M. Chopra and D.I. Thurnham. (2004). Wine polyphenols and promotion of cardiac health. *Nutrition Research Reviews,* 17, 111-129.

Costin, J. W., N. W. Barnett, S. W., Lewis, D. J., McGillivery. (2003). Monitoring the total phenolic/antioxidant levels in wine using flow injection analysis with acidic potassium permanganate chemiluminescence detection. *Anal. Chim. Acta.* 499, 47-56.

Darias-Martin, J.J., O. Rodrigues, E. Diaz, R.M. Lamuela-Raventos. (2000). Effect of skin contact on the antioxidant phenolics in white wine. *Food Chemistry,* 71, 483-487.

De Beer, D., E. Joubert, W.C.A. Gelderblom, M. Manley. (2005). Changes in the Phenolic Composition and Antioxidant Activity of Pinotage, Cabernet Sauvignon, Chardonnay and Chenin blanc Wines During Bottle Ageing *Food Chem.* 90, 569-577.

Downey, M.O., N.K. Dokoozlian, M.P. Krstic. (2006). Cultural practice and environmental impacts on the flavonoid composition of grapes and wine: A review of recent research. *Am. J. Enol. Vitic.* 57, 257-268.

Dunstan, S. L., G. B. Sala-Newby, A. B. Fajardo, K. M. Taylor, and A. K. Campbell. (2000). Cloning and expression of the bioluminescent photoprotein pholasin from the bivalve mollusc Pholas Dactylus. *J. Biol. Chem.* 275, 9403.

Fernandez-Pachon, M. S., D. Villano, M.C. Garcia-Parrilla & A.M. Troncoso. (2004). Antioxidant activity of wines and relation with their polyphenolic composition. *Analytica Chimica Acta*, 513, 113-118.

Folin, O., V. Ciocalteu. (1927). On tyrosine and tryptophane determinations in proteins. *J. Biol. Chem.* 73, 627-650.

Fuhrman, B., N. Volkova, A. Suraski, M. Aviram. (2001). White wine with red winelike properties: increased extraction of grape skin polyphenols improves the antioxidant capacity of the derived white wine. *Journal of Agricultural and Food Chemistry* 49 (7), 3164-3168.

Garcia-Campana, A. M., and W. R. G. E. Baeyens. (2001). *Chemiluminescence in Analytical Chemistry*, Marcel Dekker, New york.

Giovanelli, G. (2005). Evaluation of the antioxidant activity of red wines in relationship to their phenolic content. *Italian Journal of Food Science*, 17, 381-393.

Gómez-Alonso, S., E. García-Romero, I. Hermosín-Gutiérrez. (2007). HPLC analysis of diverse grape and wine phenolics using direct injection and multidetection by DAD and fluorescence. *J. Food Comp. Anal.* 20, 618-626.

Hipler, B., and J. Knight. (2001). ABEL® Antioxidant test kit with Pholasin® for Vitamin C type antioxidants. Knight Scientific Ltd., Application note 108.

Hötzer, A. K., C. Henriquez, E. Po, S. Miranda-Rottmann, A. Aspillaga, F. Leighton, E. Lissi. (2005). Antioxidant and prooxidant effects of red wine and its fractions on Cu(II) induced LDL oxidation evaluated by absorbance and Chemiluminescence measurements. *Free radical research*. 39, 2, 175-183.

Katalinic, V., M. Milos, D. Modun, I. Music, M. Boban. (2004). Antioxidant effectiveness of selected wines in comparison with (+)-catechin. *Food Chemistry*. 86, 593-600.

King, R. E., J.A. Bomser and D.B. Min. (2006). Bioactivity of resveratrol. Comprehensive *Reviews in Food Science and Food Safety*. 5, 65-70.

Knight, J. (1997). The piddock and the immunologist. *Immunol. News*. 4, 26.

Kočar, D., M. Strlič, J. Kolar, V. S. Šelih, and B. Pihlar. (2008). Peroxide-related chemiluminescence of cellulose and its auto-absorption. *Polymer Degradation and Stability*. 93, 263-267.

Kuse, M., E. Tanaka, and T. Nishikawa. (2008). Pholasin luminescence is enhanced by addition of dehydrocoelenterazine. *Bioorganic and Medicinal Chemistry Letters*. 18, 5657-5659.

Lopez-Velez, M., F. Martinem-Martinez, C. Dell Valle-Ribes. (2003). The study of phenolic compounds as natural antioxidants in wine. *Critical Reviews in Food Science and Nutrition*. 43, 3, 233-244.

Magalhães, L.M., M. Santos, M. A. Segundo, S. Reis, J.L.F.C. Lima. (2009). Flow injection based methods for fast screening of antioxidant capacity. *Talanta*. 77, 5, 1559-1566.

Makris,D. P., E. Psarra, S. Kallithraka, P. Kefalas. (2003). The effect of polyphenolic composition as related to antioxidant capacity in white wines. *Food Res. Int.* 36, 805-814.

Malovana, S., F. J. Montelongo Garcia, J. P. Perez, and M. A. Rodriguez-Delgado. (2001). Optimisation of sample preparation for the determination of trans-resveratrol and other polyphenolic compounds in wines by high performance liquid chromatography. *Anal. Chim. Acta.* 428, 245-253.

Michelson, A. M. (1978). Purification and properties of Pholas dactylus luciferin and luciferase. In: DeLuca MA ed) *Methods in Enzymology*, Academic Press, London 57, 385.

Minussi, R.C., M. Rossi, L. Bologna, L. Cordi, D. Rotilio, G. M. Pastore. (2003). Phenolic compounds and total antioxidant potential of commercial wines. *Food Chem.* 82, 409-416.

Mozetič, B., I. Tomažič, A. Škvarč, P. Trebše. (2006). Determination of polyphenols in white grape berries cv. Rebula. *Acta chim. slov.* 53, 58-64.

Müller, T., E. V. Davies, and A. K. Campbell. (1989). Pholasin chemiluminescence detects mostly superoxide anion released from activated human neutrophils. *J. Biolumin. Chemilumin.* 3, 105.

Nave, F., M. João Cabrita, C. T. Da Costa. (2007). Use of solid-supported liquid-liquid extraction in the analysis of polyphenols in wine. *J. Chromatogr. A.* 1169, 23-30.

Opie, L. H. and S. Lecour. (2007). The red wine hypothesis: from concepts to protective signalling molecules. *Eur. Heart J.* 28, 14, 1683-1693.

Paixao, N., R. Perestrelo, J.C. Marques & J.S. Camara. (2007). Relationship between antioxidant capacity and total phenolic content of red, rose and white wines. *Food Chemistry*, 105, 204-214.

Prior, R. L., X. Wu, and K. Schaich. (2005). Standardized Methods for the Determination of Antioxidant Capacity and Phenolics in Foods and Dietary Supplements. *J. Agric. Food Chem.* 53, 4290-4302.

Prosen, H., D. Kočar, M. Strlič, and D. Rusjan. (2007). In vino veritas: LC-MS in wine analysis. *LC-GC Eur.* 20, 617-621.

Rastija, V., G. Srečnik & M. Medić-Šarić. (2009). Polyphenolic composition of Croatian wines with different geographical origins. *Food Chemistry*, 115, 54-60.

Recamales, A. F., A. Sayago, M. L. González-Miret, D. Hernanz. (2006). The effect of time and storage conditions on the phenolic composition and colour of white wine. *Food Res. Int.* 39, 220-229.

Reichl, S., J. Arnhold, J. Knight, J. Schiller, and K. Arnold. (2000). Reactions of Pholasin with peroxidases and hypochlorous acid. *Free Radical Biol. Med.* 28, 1555.

Roberts, P. A., J. Knight, A. K. Campbell. (1987). Pholasin: a bioluminescent indicator for detecting activation of single neutrophils. *Anal. Biochem.* 160, 139-148.

Rodríguez-Delgado, M.A., S. Malovaná, J. P. Pérez, T. Borges, F. J. García Montelongo. (2001). HPLC-analysis of polyphenolic compounds in spanish red Wines and determination of their antioxidant activity by Radical scavenging assay. *J. Chromatogr. A* 912, 249-25.

Russo, G. L. (2007). Ins and outs of dietary phytochemicals in cancer chemoprevention. *Biochemical Pharmacology.* 74, 533-544.

Santos-Buelga, C. And A. Scalbert. (2000). Proanthocyanidins and tannin-like compounds-nature, occurrence, dietary intake and effects on nutrition and health. *Journal of the Science of Food and Agriculture*. 80, 1094-1117.

Shimomura, O. (2006). *Bioluminescence: chemical principles and methods*. World Scientific Publishing, New Jersey, USA.

Singleton, V. L., and J. A. Rossi. (1965). Colorimetry of total phenolics with phosphomolybdic-phosphotungstic acid reagents. *Am. J. Enol. Vitic.* 16, 144.

Spigno, G., D. M. De Faveri, (2007). Assessment of process conditions on the extraction of antioxidants from grape marc. *J. Food Eng.* 78, 793-801.

Spranger, I.M., C. M. Clímaco, B. Sun, N. Eiriz, C. Fortunato, A. Nunes, C. M. Leandro, M. L. Avelar, P. A. Belchior. (2004). Total polyphenolic compounds contents (TPC), total antioxidant activities (TAA) and HPLC determination of individual polyphenolic compounds in selected Moravian and Austrian wines. *Anal. Chim. Acta*. 513, 151-161.

Staško, A., V. Brezova, M. Mazur, M. Čertik, M. Kalinak & G. Gescheidt. (2008). A comparative study on the antioxidant properties of Slovakian and Austrian wines. *LWT- Food Science and Technology*, 41, 2126-2135.

Strlič, M., T. Radovič, J. Kolar, and B. Pihlar. (2002). Anti- and pro-oxidative properties of gallic acid in Fenton-like systems. *J. Agric. Food Chem.* 50, 6313-6317.

Tsao, R., Z. Deng. (2004). Separation procedures for naturally occurring antioxidant phytochemicals. *J. Chrom. B* 812, 85-99.

Urbano-Cuadrado, M., M.D. Luque de Castro, P.M. Pérez-Juan, J. García-Olmo, M. A. Gómez-Nieto. (2004). Near infrared reflectance spectroscopy and multivariate analysis in enology Determination or screening of fifteen parameters in different types of wines. *Anal. Chim. Acta*. 527, 81-88.

Vitrac, X., J. P. Monti, J. Vercauteren, G. Deffieux, J. M. Mérillon. (2002). Direct liquid chromatographic analysis of resveratrol derivatives and flavanonols in wines with absorbance and fluorescence detection. *Anal. Chim. Acta*. 458, 103-110.

Vršič, S., B. Pulko, J. Valdhuber. (2009). Influence of defoliation on carbohydrate reserves of young grapevines in the nursery. *European journal of horticultural science*. 74, 5, 218-222.

Weingerl V, M. Strlič, D. Kočar. (2009). Comparison of methods for determination of polyphenols in wine by HPLC-UV/VIS, LC/MS/MS and spectrophotometry. *Acta Chim. Slov.* 56, 3, 698-703.

Weingerl, V., M. Strlič, D. Kočar. (2011). Evaluation of the chemiluminometric method for determination of polyphenols in wine. *Analytical Letters*. 44, 1310-1322.

Woraratphoka, J., K. O. Intarapichet, K. Indrapichate. (2007). Phenolic compounds and antioxidative properties of selected wines from the northeast of Thailand. *Food Chem*. 104, 1485-1490.

Zorn, M. E., R. D. Gibbons, and W. C. Sonzogni. 1997. Weighted least squares approach to calculating limits of detection and quantification by modeling variability as a function of concentration. *Anal. Chem.* 69 (15), 3069-3075.

5

A Review of Spectrophotometric and Chromatographic Methods and Sample Preparation Procedures for Determination of Iodine in Miscellaneous Matrices

Anna Błażewicz

Department of Analytical Chemistry, Medical University of Lublin,
Poland

1. Introduction

Why nearly 200 years after the accidental discovery of natural iodine by Bernard Courtois (Dijon, France, 1811) are researchers still intrigued by this element? Why do they constantly search for more sensitive and reliable methods of its determination? During the last 200 years the status of research into the role of iodine in living organisms and the environment has progressed through many phases, from rapture over its interesting properties and healing powers to even some kind of "iodophobia". According to the World Health Organization (WHO), an estimated 2 billion people, including 285 million school-age children, are iodine deficient. And among them, iodine deficiency disorders (IDD) affect some 740 million -- with almost 50 million of them suffering from some form of brain damage resulting from iodine deficiency (United Nations Administrative Committee on Coordination/ Sub-Committee on Nutrition, 2000). On the other hand, it is known that large amounts of iodine are able to block the thyroid's ability to make hormones and worsen infiltration of the thyroid by lymphocytes. According to Teng's studies (Teng et al., 2009) giving iodine to people who had adequate or excessive iodine intake increases the incidence of autoimmune thyroiditis.

It became evident that increasing familiarity with the role of iodine would translate into development and discovery of more and more powerful analytical methods and techniques. Present-day analytical techniques are capable of detecting extremely small quantities. Some of them have become routine ultra-trace measurement tools in analytical and clinical laboratories.

The aim of this review is to explore available information regarding iodine determination in various samples (mainly of biological and environmental origin) focusing on spectrophotometry and chromatography as sensitive and reliable analytical methods of its measurement.

2. Iodine species in nature

Iodine plays an integral role in a diverse array of processes. As such, it exists in a variety of forms reflecting either the environment in which it is found or its biological function. Water, air, soil and food constitute the most common group of analyzed samples derived from our external environment.

2.1 Water

In water iodine is predominantly found in the iodide (I-) or iodate (IO_3^-) form (Gilfedder et al., 2007; Schwehr & Santschi, 2003). Other forms of iodine species in water include: IO_4^- (periodate), IO-(hypoiodite), CH_3I (methyl iodide), CH_2I_2 (methyl diiodide), C_2H_5I (ethyl iodide), C_3H_7I (propyl iodide), C_4H_9I (butyl iodide), and CH_2BrI (methyl iodide bromide) (Hou, 1999, Hou et al., 2009). Organic iodine concentrations may be especially high in fresh water (from rivers, lakes and rain).

2.2 Air

Generally, the air contains iodine in particulate form, as inorganic gaseous iodine (I_2, HIO) and as some forms of organic gaseous iodine (CH_3I, CH_2I_2). High iodine concentrations are found in urban areas due to the combustion of oil and coal. Coastal areas also have high iodine concentration due to the emission of gaseous I_2 from algae, seawater and sea spray, which varies with location, season and climate (Hou, 2009; Yoshida & Muramatsu, 1995). There are numerous pathways that may be responsible for transferring I_2 from the sea to air. Photochemical oxidation of I- from seawater to elemental iodine has been recreated in the laboratory (Miyake & Tsunogai, 1963). Another means of obtaining elemental I_2 is through the reaction of I- with ozone (O_3)(Garland & Curtis, 1981). The greatest source of atmospheric I_2 is thought to be from microbial activity within the oceans, through the transformation of I- and IO_3^- into organic CH_3I, which has a residence time between 1.1-8 days (Cicerone, 1981). Lovelock et al. (Lovelock et al., 1973) measured the mean atmospheric CH_3I concentration above the Atlantic as 1.2 ppt (6.8 ng/m³). Rasmussen et al. (Rasmussen et al., 1982) found the background level of CH_3I to vary between 1 and 3 ppt (5.7-17 ng/m³) with measurements near oceans with high biomass productivity to be around 7-22 ppt (40-125 ng/m³).

2.3 Soil

Strong evidence suggests that atmospheric transport from the oceans is responsible for the deposition of iodine in soil. Iodine exists in various forms in soil and varies largely with respect to its concentration. I- (more mobile form) is believed to be the dominant species in acidic soils whilst IO_3^- (less mobile) will occur in alkaline soils. In low pH oxidizers, e.g. Fe^{3+} and Mn^{4+} convert I - into molecular I_2. The activity of reducing bacteria impacts the form of iodine in the soil as well. CH_2I_2 and other volatile organic complexes of iodine are generated by microbial activity (Johnson, C.C. 2003). Generally, organic bound iodine is more abundant in soil samples. The secondary environment (soil) has high iodine content compared to the primary environment (parent rocks) from which it is derived as a result of weathering. Weathered rocks and soils are richer in iodine than the unweathered bedrocks (Fuge & Johnson, 1986). On average, the igneous rocks contain an average of 0.25 mg/kg of iodine, the sedimentary rocks have 2.3 mg /kg iodine and the metamorphic rocks have 0.81 mg/kg of iodine. Organic matter is the major concentrator of iodine in sedimentary basins (Mani et al., 2007). The highest values were found in soil samples from areas close to the coast, where there is high rainfall, and from areas with high soil organic matter. In one study, the iodine concentration in Japanese soils was found to range from 0.2 mg/kg to 150

A Review of Spectrophotometric and Chromatographic Methods and Sample Preparation Procedures for Determination
of Iodine in Miscellaneous Matrices

83

mg/kg (Muramatsu, 2004). Retention of iodine in the soil is influenced by a number of factors, including the soil pH, moisture content, porosity and composition of the organic and inorganic components (Sheppard et al., 1995; Whitehead, 1984).

2.4 Food and plants

Geographic differences in topsoil iodine content and irrigation procedures determine food iodine levels. In most diets the mainstay sources of iodine are fish, shellfish, milk and iodinated salt. Food supplements constitute an alternative means of obtaining dietary iodine. Fish contain iodine in similar forms to those found in humans. Perhaps the widest assortment of forms is encountered in different species of seaweed. Brown seaweeds contain mostly I-, however green seaweeds play host to a wide array of organic molecules to which iodine is bound, including numerous proteins and polyphenols. Dietary iodine is obtained from a variety of sources and individual dietary habits contribute to the wide disparity in iodine intake among populations. In 1993, the World Health Organization [WHO] published the first version of the WHO Global Database on Iodine Deficiency with global estimates on the prevalence of iodine deficiency based on total goitre prevalence (TGP), using data from 121 countries. Since the international community and the authorities in most countries where IDD was identified as a public health problem have taken measures to control iodine deficiency, in particular through salt iodization programmes – the WHO recommended strategy to prevent and control IDD (World Health Organization [WHO], 2004). Salt iodization programmes are carried out in more than seventy countries, including the United States of America and Canada. There is a wide variation in the scope of iodine supplement; almost 90% of households in North and South America utilize iodized salt while in Europe and the East Mediterranean regions this figure is less than 50%, with a worldwide figure about 70% (WHO, 2007). WHO, ICCIDD (International Council for the Control of Iodine Deficiency Disorders) and UNICEF (United Nations International Children's Emergency Fund) recommend that the term "iodized" be used to designate the addition of iodine to any substance, regardless of the form. Iodine is commonly added as the I- or IO_3^- of potassium, calcium or sodium. 70% of salt sold for household use in the U.S.A. is iodized with 100 ppm KI (400 μg iodine per teaspoon) (U.S. Salt Institute, 2007). In Canada all salt must be iodized with 77 ppm KI. Mexico requires 20 ppm levels of iodization. Recommendations for the maximum and minimum levels of iodization of salt are calculated as iodine and determined by local national health authorities in accordance with regional variations in iodine deficiency. In Poland iodine deficiency prophylaxis was first started in 1935. Near the end of the 1940's and 80's, the practice of iodizing table salt was abandoned. A nonobligatory recommendation of iodizing salt took place in 1989. In 1991 the Polish Council for Control of Iodine Deficiency Disorders (PCCIDD) was established and an epidemiological survey performed in 1992-1993, defined Poland as an area with moderated – the seaside region as light – severity of iodine deficiency. In 1996 the production of table salt without the addition of KI was made illegal. An obligatory law was passed, mandating the addition of 30(+/-10) mg of KI per kilogram of salt (Szybiński, 2009). Iodization of salt in Turkey has been mandatory since 1998 and the recommended iodine concentration is 50-70 mg KI/kg or 25-40 mg KIO₃/kg (Gurkan et al., 2004). In Switzerland, iodization of salt was altered three times. Iodization was first introduced in 1922 at 3.75 mg/kg. In 1962 the concentration was doubled and in 1980 it was doubled again giving a present level of 15 mg/kg. Although it's use is voluntary, by 1988, 92% of retail salt and 76% of all salt for human consumption (including food industry) was iodized. Most other countries add from 10 to 40 µg iodine per gram of salt (10-40 ppm) (Bürgi et al., 1990).

Bread (0.14 mg/kg), milk (0.32 mg/kg), eggs (0.48 mg/kg), meat (0.13 mg/kg), and poultry (0.1 mg/kg) constitute other important sources of iodine (figures in parentheses represent the average iodine content per fresh weight) (Food Standards Agency [FSA], 2000). In certain individuals, medications may contribute to the ingested daily iodine. Examples include amiodarone, an antiarrythmic agent (Fang et al, 2004), iodized intravenous radiographic contrast agents and certain topical antiseptics (Aiba et al., 1999).

When considering multivitamins and mineral supplements as a source of iodine, one can find that the majority of iodine they contain is in the KI or NaI forms. According to Zimmerman's research, iodine concentrations in plant matter can range from as little as 10 µg/kg to 1 mg/kg dry weight (Zimmermann, 2009). This variability is relevant because plant matter affects the iodine content of meat and animal products (Pennington et al., 1995). Iodine content of different seaweed species varies greatly (Teas et al., 2004). Japanese iodine intake from edible seaweeds is relatively high compared to the rest of the world. Having taken into consideration many factors, such as information from dietary records, food surveys, urinalysis and seaweed iodine content, Zava and Zava estimated that the daily iodine intake in Japan averages approximately 1,000 to 3,000 µg/day (Zava & Zava, 2011). In certain diets, seafood is a large source of iodine, containing 2 to 10 times more iodine than meat (Hemken, 1979). Saltwater seafood usually contains significantly more iodine than freshwater food, some edible seaweeds may contain up to 2500 µg iodine per gram (Teas et al., 2004).

Simon et al. (Simon et al., 2002) presented an example of the value of the determination of iodine compounds in fish. The authors analyzed whole-body homogenates of zebrafish (Danio rerio) and tadpoles of the African clawed frog (Xenopus laevis). They detected five previously unknown iodinated compounds and measured the concentrations of I-, MIT, DIT, T4, T3 and rT3 in these species.

2.5 Human body

In relation to iodine determinations, in clinical practice the most frequent analytical samples include urine, serum, blood, and a variety of tissues. Therefore, some examples of research studies related to iodine determinations in the mentioned matrices are presented below. The bioavailability of organic iodine, especially associated with macromolecules, is low (Hou et al., 2009), whereas I- and IO_3- have high bioavailability. According to recent estimates, KI is almost completely absorbed in humans (96.4%) (U.S. Food and Drug Administration [FDA], 2009).

2.5.1 Thyroid

Iodine plays a key structural role in the thyroid hormones of humans and other mammals, primarily in the form of T3 (triiodothyronine) and T4 (thyroxine). In such samples precursor forms such as MIT (monoiodotyrosine) and DIT (diiodotyrosine) or isomer forms such as rT3 (reverse triiodothyronine) may also be measured. Iodine accounts for 65% of the molecular weight of T4 and 59% of the T3. 15–20 mg of iodine is concentrated in the thyroid and hormones with 70% distributed in other tissues. In the cells of these tissues, iodide enters via the sodium-iodide symporter (NIS).

A Review of Spectrophotometric and Chromatographic Methods and Sample Preparation Procedures for Determination of Iodine in Miscellaneous Matrices

85

According to Hou (Hou et al. 1997), the contents of iodine expressed as ng/g wet weight tissue±1SD) in five tissues, plus hair, averaged over 9–11 individuals were: the heart (46.6±14.9), liver (170±34), spleen (26±8.6), lung (33.3±10.6), muscle (23.5±14.3), and hair (927±528). In the U.S. population, Okerlund found a mean value of 10 mg iodine per thyroid, with a range of 4-19 mg. In 56 patients suffering from autoimmune thyroiditis but with normal thyroid function, a mean value of 4.8 mg/thyroid was reported. In 13 patients with autoimmune thyroiditis and hypothyroidism, the mean value was 2.3 mg/thyroid (Okerlund, 1997).

Zaichick and Zaichick (Zaichick & Zaichick, 1997) used instrumental neutron activation and X-ray fluorescent analyses to determine the concentration and total iodine content of iodine within thyroids. They obtained 90 samples (at autopsy) from subjects of a broad age spectrum, from 2 to 87 years old and calculated correlations between iodine concentration and age. All their thyroid samples were weighed, lyophilised and homogenised. Iodine was analyzed in approximately 50-mg samples. The mean intrathyroidal iodine concentration (mean +/- S.E.) of a normal subject aged 26-65 averaged 345 +/- 21 µg/g dry tissue in non-endemic goitre region with no obligatory salt iodination. Maximum iodine concentration was found to be 494 +/- 65 µg/g (P < 0.05) for the age of 16-25. For the elderly aged over 65 an increase in iodine of 668 +/- 60 µg/g was shown (P < 0.001). When comparing the right and left lobes, the authors found no variation in weight, iodine concentration or the total content. An inverse correlation was found between the thyroid weight and intrathyroidal iodine concentration (-0.32, P < 0.01).

Tadros et al. (Tadros et al., 1981) determined iodine in 48 normal thyroids obtained at autopsy. According to the authors' findings, the iodine concentration ranged from 0.02 to 3.12 mg/g of tissue with a mean value of 1.03 +/- 0.67 mg/g. In 91 surgical thyroid specimens with a variety of abnormalities they found that iodine concentration was much lower. The samples of thyroids with cancer had the lowest values. Sixteen (76%) of 21 analyzed malignant thyroid specimens had undetectable iodine (less than 0.02 mg/g), whereas 22 (96%) of 23 benign nodules had measurable iodine concentrations. Błażewicz et al. (Błażewicz et al., 2011) examined correlations between the content of iodides in 66 nodular goitres and 100 healthy human thyroid tissues. The authors presented an accurate assessment of the iodine content in the thyroids of patients with a nodular goitre (mean concentration was 77. 1 +/- 14.02 µg/g) and in the thyroids obtained at autopsy - considered as a control group (mean concentration 622.62 +/- 187.11 µg/g -for frozen samples and 601.49 +/- 192.11 µg/g- for formalin fixed samples). Statistical analysis showed approx. 8-fold reduction of iodine concentration in the pathological tissues in comparison with the control group.

Interesting research into iodine content in human thyroids was also conducted by Zabala et al. (Zabala et al., 2009). Their study focuses on the determination of iodine content in healthy thyroid samples on male population from Caracas in Venezuela. The authors aimed at establishing a baseline of iodine content in thyroid glands and hence to compare the iodine thyroid concentration of the Venezuelan population with other countries. Male post-mortem individual samples were analyzed using a spectrophotometric flow injection method, based on the Sandell-Kolthoff reaction. The median intrathyroidal iodine concentration was 1443+/-677 µg/g (wet weight), ranging from 419 to 3430 µg/g, which corresponds to a median of total iodine content of 15+/-8 mg (ranging from 4 to 37). These results were higher than those values which were found in the literature. No correlation of iodine content with the age or weight of the healthy gland was observed.

2.5.2 Plasma

The inorganic form of iodine represents about 0.5 % of the total plasma iodine. The rest occurs in bound form with specific plasma protein (protein - bound iodine, PBI) which has gained wide use as an indicator of thyroid activity in humans. It has been reported that the total plasma iodine concentration in healthy subjects is between 40 and 80 µg/l. According to Allain's studies when plasma iodine concentrations are below 40 µg/l, hypothyroidism is highly likely, when they are between 80 and 250 µg/l, hyperthyroidism, particularly Graves' disease is probable. Above 250 µg/l – iodine overload is almost certainly indicated (Allain et al., 1993).

2.5.3 Brain

Despite the fact that iodine is one of the most important essential elements, the quantitative data on its concentration in the human brain is really scarce. The nature and site of iodine binding in the human brain is still unknown. The results of Andrasi et al. (Andrási et al., 2004) investigations on iodine distribution between the lipid fraction and in the brain tissue without lipid have indicated that its mean contents vary between 910 ± 147 ng/g dry weight and 281 ± 68 ng/g dry weight depending on the brain region (the highest content was found in susbstantia nigra and the lowest in vermis crebelli).

2.5.4 Hair

Levine et al. (Levine et al., 2007) presented a study on determining iodine concentration in tiny (less than 25 mg) human hair samples. Iodine concentrations from the blinded hair autism study samples ranged from 0.483 to 15.9 µg/g. In Adams' et al. studies (Adams et al., 2006) the mean concentration of iodine in the hair of autistic children has been reported to be lower than in the hair from the control group children. The low level of iodine in the hair of children with autism suggests that iodine could be important in the aetiology of autism, presumably due to its effect on thyroid function.

2.5.5 Human milk

Approximately 80% of iodine in human milk is present as I-, while, mainly, another six high molecular weight iodine containing molecules (Braetter et al., 1998) account for the remaining 20%. European breast milk samples have been determined to contain 95 ± 60 µg/l total iodine. The total iodine content varied depending on the lactation state, and iodine was associated with fat at approximately 30% and 70% of the low molecular weight fraction (Michalke, 2006). A study of iodine species in milk samples obtained from humans from several different European countries and in infant formulas from different manufacturers was carried out by Fernández-Sanchez and Szpunar (Fernández-Sanchez & Szpunar, 1999). The authors also developed a method to determine iodine in human milk and infant formulas using ICP-MS. In the human milk the values found were between 144 ± 93.2 µg/kg, whereas in the infant formulas the values were 53.3 ± 19.5 (Fernández-Sánchez et al., 2007).

A Review of Spectrophotometric and Chromatographic Methods and Sample Preparation Procedures for Determination of Iodine in Miscellaneous Matrices

87

2.5.6 Urine

In urine, iodine occurs as I⁻, but some organic species can also be found. Urinary iodine concentration is the prime indicator of nutritional iodine status and is used to evaluate population-based iodine supplementation. In 1994, WHO, UNICEF and ICCIDD recommended median urinary iodine concentrations for populations of 100- 200 µg/l, assuming the 100 µg/l threshold would limit concentrations <50 µg/l to </=20% of people (Delange et al., 2002). During the period between the years 1994-2002, the urinary iodine concentration was determined in 29,612 samples at the Institute of Endocrinology in the Czech Republic. The mean basal urinary iodine concentrations +/-SD were 115+/-69 µg/l. Out of all the samples, 0.7% were in severe (<20 µg/l), 9.6% in moderate (20-49 µg/l), 40.1% in mild (50-99 µg/l), 35.6% in adequate (100-200 µg/l), and 14.0% in more than adequate (>200 µg/l) subsets of iodine nutrition. A statistically significant (p<0.00001) difference was found between the mean male (127 µg/l) and female (112 µg/l) urinary iodine, and an inversely proportional trend also existed in the age-related data (Bílek et al., 2005). It is also known that patients with iodine induced hyperthyrosis have 10- to 100-fold more urinary iodide than healthy patients (Mura et al., 1995).

Delange et al. (Delange et al., 2002) determined the frequency distribution of urinary iodine in iodine-replete populations (schoolchildren and adults) and the proportion of concentrations <50 µg/l. The findings were as follows: nineteen groups reported data from 48 populations with median urinary iodine concentrations >100 µg/l. The total population was 55 892, including 35 661 (64%) schoolchildren. Median urinary iodine concentrations were 111-540 (median 201) µg/l for all populations, 100-199 µg/l in 23 (48%) populations and >/=200 µg/l in 25 (52%). The frequencies of values <50 µg/l were 0-20.8 (mean 4.8%) overall and 7.2% and 2.5% in populations with medians of 100-199 µg/l and >200 µg/l, respectively. The frequency reached 20% only in two places where iodine had been supplemented for <2 years. According to the authors' conclusions the frequency of urinary iodine concentrations <50 µg/l in populations with median urinary iodine concentrations >/=100 µg/l has been overestimated, and the threshold of 100 µg/l does not need to be increased. The main conclusion of the cited work was that in populations, median urinary iodine concentrations of 100-200 µg/l indicate adequate iodine intake and optimal iodine nutrition.

According to Verheesen and Schweitzer (Verheesen and Schweitzer, 2011) the threshold of 100 µg/l is only to make sure that severe iodine deficiency (beneath 50 µg/l) is not present in more than 20% of the population. Although the WHO is concerned about the negative effects of even mild iodine deficiency, the 100 µg/l threshold was never intended to prevent mild iodine deficiency. In order to combat mild deficiency the threshold should be reconsidered. The authors also emphasized the need to test for other biomarkers in individual cases in order to be able to adequately establish iodine deficiency. Since there is a lack of trusted biomarkers, thus far statistics have been used to estimate the percentage of the population being deficient, instead of showing prevalence figures. Furthermore, population figures are typically described only by a median; variables such as % being deficient, % being pregnant, % women, % men and age related figures should be thoroughly investigated.

3. Problems with analysis of iodine in biological matrices

Despite a wide choice of available analytical methodologies, determination of iodide in biological matrices remains a difficult problem. Biological samples belong to so-called complex samples (with complex matrices). In such samples, the analyte content is usually much scarcer when compared with the accompanying macrocomponents. Aside from the necessity of choosing the appropriately sensitive method, it is equally important to comply with the sample preservation, pretreatment, and preparation conditions.

Historically, published values of the I_2/I^- concentration of both tissue and body fluids from healthy subjects have varied greatly. These great differences were attributed to numerous variables, such as age, sex, dietary habits, physiological conditions, environmental factors and numerous other X-factors. Given the delicate nature and the instability of biological samples, it has been concluded that improper sample collection methods and manipulation drastically affects the iodine content of biological matrices.

Analytical methods are often versatile in nature. Thus, in order to achieve successful and satisfactory results, the process of analysis needs to be carefully tailored to its needs. Before applying the appropriate method for a particular application, many factors have to be considered and some of them are discussed below.

It is well known that sources of errors that affect the final error of an analytical result are connected, among other things, with incorrect obtaining of the samples, their improper transport, storage and transformation, wrong methodology, wrong measurement (instruments, parameters) or human errors (Konieczka & Namieśnik, 2007). When it comes to quantitative evaluations of iodine concentrations (in all chemical forms), the proper storage of biological samples is of paramount importance. The tissues must be preserved in such a way that a potential loss of the analyte (i.e. iodine) is minimized. The choice of the tissue-fixing agents is quite wide. Formalin is a routinely used tissue-fixing agent after surgical procedures. Other recommended agents for tissue preservation include, e.g. a mixture of 50% glutaraldehyde, 16 % paraformaldehyde, and 0.2M sodium phosphate buffer solutions (i.e. original composition of Karnovsky fixative) or its modification. The other possibility to preserve the tissues is sample freezing.

Hansson at al. (Hansson et al., 2008) used X-ray fluorescence analysis (XRF) and secondary ion mass spectrometry (SIMS) for evaluation of a freezing technique for preserving samples (XRF analysis) and for evaluation of the efficacy of using aldehyde fixatives to prepare samples (SIMS analysis). There were no significant changes in the iodine content due to freezing. Freezing for 4 weeks produced no more than a 10% change in the iodine content. For all the samples fixed in an aldehyde, there was a loss of iodine. The decrease in iodine content from baseline was significant for samples fixed in aldehyde ($p < 0.05$). Karnovsky was the best fixative in this regard, yielding a mean 14% loss compared to 20% and 30% for glutaraldehyde and formaldehyde, respectively. For SIMS method, Fragu et al. (Fragu et al., 1992) recommended chemical fixation with a mixture of Karnovsky fixative, followed by embedding in methacrylate. This method, which was evaluated for iodine loss by Rognoni et al. (Rognoni et al., 1974), has proven suitable for preservation of substances bound to macromolecules (like iodine bound to thyroglobulin [Tg]).

The effect of sample preservation on determination of I^- in healthy and pathological human thyroids has also been studied (Błażewicz et al., 2011). It was pointed out that the way of tissue preservation (either in formalin or by freezing) had no significant effect on the iodine determination result ($\alpha = 0.1$) by ion chromatography combined with the pulsed amperometic detection method (IC-PAD). Sample decomposition is a critical step in iodides' analysis as well. All reported methods have a digestion or ashing step prior to the final determination of

A Review of Spectrophotometric and Chromatographic Methods and Sample Preparation Procedures for Determination
of Iodine in Miscellaneous Matrices

89

iodine. The procedure usually requires the use of alkaline media and a high temperature (Moxon & Dixon, 1980). A catalytic spectrophotometric method based on the Sandell-Kolthoff reaction together with many modifications and improvements of this method is a low-cost assay of iodine, however, it is not free of possible interferences (especially for foodstuffs with low range of iodine levels). Most iodine in biological media is covalently bonded and there are some substances that interfere with the determination reaction (eg. SCN^-, NO_3^-, or Fe^{2+}). A high lipid content in the sample (eg. milk) can cause problems in spectrophotometric readings, so a mineralization step is absolutely necessary before analysis.

When preparing a sample for analysis, it is necessary to take into consideration also the loss of analyte due to erroneous application of decomposition procedures. Some of them, e.g. "Schoeniger combustion", require a highly homogenous sample, which is sometimes difficult to obtain (Knapp et al., 1998). There are many options for biological sample preparations, among which alkaline digestion using tetramethylammonium hydroxide (TMAH) is the most common before the analysis of I^- (Fecher et al., 1998; Fecher & Nagengast, 1994; Schramel & Hasse, 1994). Alkaline conditions during the extraction procedure have some advantages in comparison with acidic media, where I^- may be oxidized into volatile forms (I_2 or HI). A destruction of the organic matrix using a typical digestion reagent, like HNO_3, is not possible because of the losses due to volatile iodine formation (therefore, no stable sample solution can be achieved). However, acid digestion procedures (by the use of mainly H_2SO_4, HNO_3, and $HClO_4$) have also been applied (Fischer et al., 1986). What is more, despite its obvious weak point (loss of analyte), the US Food and Drug Administration (FDA, 2009) still recommends such procedures.

Błażewicz et al. (Błażewicz et al., 2011) used the alkaline digestion with 25 % TMAH water solution for the thyroid glands` preparation before the IC method of analysis. A diluted TMAH solution has also been used for serum samples by Schramel and Hasse (Schramel & Hasse, 1994) to analyse iodine in the serum, milk, plants and tissues by using the ICP-MS method.

Unfortunately, despite the abovementioned advantages, the use of TMAH solution for the digestion of samples has some disadvantages as well. Since for all modern sample pretreatment methods time is a very important factor, the long procedure time still remains a huge problem. As reported in the literature, digestion of biological materials with TMAH usually requires up to 6 hours (Gamallo-Lorenzo et al., 2005). However, the assistance of microwaves significantly shortens the time of the sample preparation step (less than 20 minutes). Such microwave-assisted alkaline digestion has been developed before the IC analysis of thyroids' samples (Błażewicz et al., 2011). It is important to monitor all conditions of the digestion procedure, especially temperature, in order to avoid the decomposition of TMAH and bursting of the closed vessels (therefore a temperature lower than 100 °C is recommended). Each time the vessels must be thoroughly cleaned with a digestion mixture in order to avoid memory effects (adsorption by the walls of containers) and consequently the loss of analyte.

4. Spectrophotometry and chromatography as tools for iodine assessment in miscellaneous matrices

It is known that the choice of the proper analytical method depends on the intended application, the number of samples, the cost of analysis and the technical capability.

Currently, there are multiple distinctive analytical methods for determining concentrations of iodine species. The methods vary in principle, reliability, accuracy, precision, availability, detection limit, sample throughput, time and reagent consumption, ease of performance and cost of analysis. These factors all play a role in the choice of the most suitable method but ultimately the purpose of the analysis determines the method, e.g., whether the analysis is routine or if an analysis of a reference material is necessary. For any given purpose, one of the first factors taken into account is whether the method's detection limit is adequately low. Several methods of iodine determination have been proposed, including catalytic methods (with LOD=0.1 µg/ l) (Kamavisdar & Patel, 2002), chromatography in various modes (eg., IC with LOD = 0.1-0.8 µg/l (Hu et al., 1999; Bichsel & Von-Gunten, 1999), (chromatographic methods are especially useful for iodine speciation when coupled with ICP-MS or elecrochemical detection), GC-EC: gas chromatography with electron capture detection (0.11µg/l) (Maros et al., 1989), GC–MS: gas chromatography–mass spectrometry(0.010 µg/l) (Das et al., 2004), FAAS: flame atomic absorption spectrometry (2.75 µg/l) (Yebra & Bollaín, 2010), NAA (0.1-0.2 µg/l) (Hou et al., 1999), ETAAS: electrothermal atomic absorption spectrometry (1.2 -3.7 µg/l) (Bermejo-Barrera et al., 1999), inductively coupled plasma mass spectrometry ICP-MS (1.0-9.0 µg/l) (Fernandez-Sanchez & Szpunar 1999), ICP-AES (40.0-470.0 µg/l) (Anderson & Markowski, 2000), inductively coupled plasma optical emission spectrometry (ICP- OES) (2 µg/l) (Naozuka et al., 2003), ion selective electrodes (1.96 µg/l) (Kandhro et al., 2009), X-rayfluorescence (XRF) (180 µg/L) (Varga, 2007), VG-ICP-OES: vapour generation inductively coupled plasma optical emission spektrometry (20 µg/l) (Niedobová at al., 2005). The iodine content can also be measured by the use of titrimetric methods usually combined with potentiometric measurements. They are also used for verifying other methods (Gottardi, 1998). The titrimetric method is mainly used for samples without complex matrices (i.e. water or salt). Generally such methods involve acidification of the sample solution and adding an excess of KI solution to determine the liberated iodine by titration with sodium thiosulphate. Despite numerous advantages of the above-mentioned methods, very few of them are widely used due to very high costs of instrumentation, software, and maintenance. Spectrophotometric and chromatographic methods are used very frequently for the analysis of iodine and its various chemical forms. Chosen examples of applications of iodine determinations are presented below.

4.1 Spectrophotometric methods

4.1.1 Water samples

Spectrophotometric analysis continues to be one of the most widely used analytical techniques available. Kinetic spectrophotometric methods, which are based on the reaction, found by Sandell and Kolthoff (1934) set the foundation for the development of different methods for the determination of iodine in environmental samples (mostly water). The said reaction proceeds according to the following equation (1):

$$2\ Ce^{4+} + As^{3+} \rightarrow 2\ Ce^{3+} + As^{5+} \tag{1}$$

By adding an arsenious acid (H_3AsO_3) solution and an ammonium cerium sulfate ((NH_4)$_2$Ce(SO$_4$)$_3$) solution as reagents to I- in a specimen, yellow Ce^{4+} is reduced to produce colorless Ce^{3+} ((2) and (3)).

$$2Ce^{4+} + 2I^- \rightarrow 2Ce^{3+} + I_2 \tag{2}$$

$$I_2 + As^{3+} \rightarrow 2I^- + As^{5+} \tag{3}$$

Iodine has a catalytic effect upon the course of reaction (1), i.e., the more iodine is present in the preparation to be analyzed, the more rapidly proceeds the reaction (1). The speed of reaction is proportional to the iodine concentration. In this manner it is possible to determine iodine even in the nanogram range. Sandell and Kolthoff found that Os and Ru catalyse this reaction in the same manner as I_2, while Mn and MnO_4^- do so in the presence of Br^-. Among substances reducing Ce^{4+} they listed NO_2^-, SCN^-, Fe^{2+}, while BrO_3^-, MnO_4 were classed by them as oxidising As^{3+}. These authors also pointed out that certain substances, such as F^-, form compounds with Ce^{4+} giving a stable complex. Ag^+, CN^-, and Hg^+ react with I^- as well. The effect of various concentrations of NaCl, NaF, KH_2PO_4, $ZnSO_4$, KCl, $MgSO_4$, KBr and of $CuSO_4$ on the described reaction was studied by Stolc (Stolc, 1961). According to the author the substances studied may be grouped into two categories, i.e. reaction inhibiting (NaF, KH_2PO_4, $ZnSO_4$, KCl) and reaction stimulating agents (NaCl, $MgSO_4$, KBr, $CuSO_4$).

The method is achieved in the following manner: a measured amount of an arsenous oxide (As_2O_3) solution in concentrated H_2SO_4 is combined with the test solution. This mixture is then adjusted to its reaction temperature, usually between 20 and 60 degree C. Cerium (IV) sulfate in sulfuric acid is then added, after which the solution is able to react for a limited time at the set temperature. The reaction time ranges from 10 to 40 minutes, and subsequently the content of the test solution of cerium (IV) ions is photometrically determined. The lower the determined concentration of cerium (IV) ions, the faster the reaction, thus a larger amount of catalyzing agent, i.e., iodine. By these means it is possible to directly and quantitatively measure the iodine concentration of the test solution, though execution of such processes is complicated and demands extensive measuring times (Sandell & Kolthoff, 1934, 1937). The above-described method was modified in various ways, for example by replacing H_2SO_4 with HNO_3 (used for acidifying the reaction mixture). It was found that the catalytic activity of iodine in HNO_3 solution is 20 times that in H_2SO_4 and is also far less sensitive towards accompanying ions, making the system far more useful for the determination of traces of iodine (Knapp & Spitzy, 1969). The reaction mixture's change in composition multiplies the sensitivity of the reaction by twenty. Consequently, test solutions of an iodine content that, according to the conventional catalytic reaction method utilizing sulfuric acid, required a reaction time of approximately 20 minutes in order to display a notable decrease in cerium (IV) ion concentration need only 1 minute to produce the same result. These results were achieved while operating at the same reaction temperature. Rodriguez and Pardue (Rodriguez & Pardue, 1969) studied the effect of H_2SO_4, $HClO_4$, Ag(I), Hg(II), Cl^- and temperature on the aforemetioned kinetic reaction. Their studies utilized the catalytic action of iodide on the decomposition of the $FeSCN^{2+}$ complex ion. This indicator reaction is characterized by an induction period, the length of which depends on the reagent concentration, pH and temperature. The mentioned method was adopted as a standard method for iodide determination in natural and waste waters as well as in food and biological samples. However, high inter-laboratory relative standard deviations have frequently been reported for this method. Some authors have suggested that this might be partly attributed to the limitations of the method to quantitatively detect or tolerate IO_3^- that are found in natural waters (Heckwan, 1979).

An alternative flow injection spectrophotometric method for the determination of I- in the ground and surface water was reported by Kamavisdar and Patel (Kamavisdar & Patel, 2002). The method was based on the catalytic destruction of the colour of the Fe(III)–SCN-–CP+-$_n$BP$_y$ quarternary complex. The detection limit of the method was reported to be 0.1 ng ml^{-1} of iodide. Another redox reaction between chloramine-T and N,N'-tetramethyldiaminodiphenylmethane (Feigl's Catalytic Reaction) was applied for the determination of traces of iodine in drinking water (Jungreis & Gedalia, 1960).

An alternative to the Sandell-Kolthoff method was developed by Gurkan et al., (Gurkan et al., 2004). Iodides were determined in waters by inhibition kinetic spectrophotometric method based on the inhibitory effect of I- on the Pd(II)-catalyzed reduction of Co(III)-EDTA by the hypophosphite ion in a weak acid medium. The main advantage of this method was related to the pretreatment step of the analysis which would be omitted (a time-consuming alkaline ashing preparative procedure is necessary in order to apply the standard method). The sensitivity of the method allowed determinations in the range of 2-35 ng/ml of I– (LOD=1.2 ng/ml). Koh et al. (Koh et al., 1988) separated I- from other chemical species by its oxidation and subsequent extraction into carbon tetrachloride. The proposed spectrophotometric method was based on the extraction of the back-extracted iodide into 1,2-dichloroethane as an ion pair with methylene blue. The authors applied that method to determine various amounts of iodide in natural water samples (at the 10^{-6} mol l^{-1} level). Spectrophotometric determination of the total dissolved sulfide in natural waters allowed also simultaneous determination of other UV-absorbing ions, including I- (Guenther et al., 2001).

4.1.2 Soil and plant samples

Different analytical techniques have been developed to extract and measure iodine concentration from the soil. The reduction of Ce (IV) by As (III) catalyzed by iodine can be used to determine the low concentration of iodine in plant and soil samples. The sample preparation requires a specialized combustion apparatus and trapping systems for iodine. For plant samples and biological materials, halogen extraction using TMAH under mild conditions has proved to be effective (Knapp et al., 1998).

Kesari et al. (Kesari et al., 1998) developed a simple and sensitive spectrophotometric method for the determination of iodine in tap water, sea water, soil, iodized salt and pharmaceuticals samples. The said method was based on oxidation of I- to IO$_3$- with bromine water and liberation of free I$_2$ from IO$_3$- by addition of KI in acidic medium. I$_2$ is then reacted with leuco crystal violet and the crystal violet dye liberated shows maximum absorbance at 591 nm. Beer's law is obeyed over the concentration range from 0.04 to 0.36 mg/l of iodine in a final solution volume of 25 ml. The method is free of interference of other major toxicants.

Lu et al. (Lu et al. 2005) applied the arsenic-cerium redox method for assessment of iodine content in soils and waters. Mean iodine concentration in soil samples was found to be 1.32 ± 0.14 mg/kg, and its content correlated positively with the water iodine content. In association with the photometric analytical technique, the alkaline dry ashing method (adding KOH and ZnSO$_4$), along with digestion via the calorimetric bomb and the utilization of the Schoniger digestion arc, provide a means for obtaining reliable results. For this investigation, the influence of iodine fertilisation on the iodine concentration of cress

A Review of Spectrophotometric and Chromatographic Methods and Sample Preparation Procedures for Determination
of Iodine in Miscellaneous Matrices

93

(Lepidium sativum) was determined by an experiment in which different amounts of iodine were added to the potted plants. The iodine fertiliser used was natural caliche. The results show a very close correlation between the iodine supply and iodine concentration in the cress which increased to more than 30 mg/kg dry matter (Jopke et al., 1996).

4.1.3 Foodstuffs

A semi-automated method for determination of the total iodine in milk was described by Aumont (Aumont, 1982). The method involved destruction of organic matter by alkaline incineration and automated spectrophometric determination of iodide based on the Sandell and Kolthoff's reaction. The recoveries of the added iodide before calcination were between 90.05 +/- 7.36% and 97.14 +/- 4.56% (mean +/- S.D.). The coefficient of variation ranged from 2.15 to 7.21% depending on the iodine content in the milk. The limit of detection was estimated to be around 2 µg/kg.

The iodide-catalyzed reaction between As(III) and Ce(IV) stopped by the addition of diphenylamine-4-sulfonic acid was used for the development of a sensitive kinetic procedure for determining iodides with a detection limit of 2 ng/mL. The developed procedure was suitable for the determination of the total iodine in foodstuffs (Trokhimenko & Zaitsev, 2004).

Another modification of the catalytic kinetic spectrophotometric method has been established for the determination of iodine using the principle that potassium periodate oxidize rhodamine B (RhB) to discolor and I^- has a catalytic effect on the reaction. The absorbance difference (ΔA) is linearly related with the concentration of iodine in the range of 0 – 2.6 µg/mL and fits the equation $\Delta A = 0.1578\ C(C:\ \mu g/mL) + 0.0052$, with a regression coefficient of 0.9965. The detection limit of the method is 7.10 ng/mL. The method was used to determine iodine in kelp, potato, tap water, and rain water samples. The relative standard deviation of 13 replicate determinations was 1.81–2.10%. The recovery of the standard addition of the method was 96.2–99.2% (Zhai et al., 2010).

Some researchers reported that the spectrophotometric methods for the determination of IO_3^- are based on its reaction with the excess I^- to liberate I_2 which forms tri iodide (Afkhami & Zarei 2001; Ensafi & Dehaghi, 2000).

Balasubramanian and Nagaraja (Balasubramanian & Nagaraja, 2008) described a sensitive spectrophotometric method for the determination of multiple iodine species such as I^-, I_2, IO_3^- and IO_4^-. The method involved oxidation of iodide to ICl_2^- in the presence of iodate and chloride in an acidic medium. The formed ICl_2^- bleaches the dye methyl red. The decrease in the intensity of the colour of the dye is measured at 520 nm. Beer's law is obeyed in the concentration range 0-3.5 µg of iodide in an overall volume of 10 ml. The relative standard deviation was 3.6% (n=10) at 2 µg of iodide. The developed method can be applied to the samples containing iodine, iodate and periodate by prereduction to iodide using $Zn/H(^+)$ or $NH_2NH_2/H(^+)$. The effect of interfering ions on the determination was pointed out. The described method was successfully applied to determine iodide and iodate in salt samples and iodine in pharmaceutical preparations.

Silva et al. (Silva et al., 1998) outlined a new method for the determination of iodate in table salt. KIO_3, after being converted to I_3 by reacting with iodide in the presence of phosphoric acid, was spectrophotometrically determined at two well defined UV absorption maxima of 352 and 288 nm. The results were comparable with a standard, ranging from 37.39 (±0.15) to 63.67 (±0.16) mg KIO_3 per kg of salt with samples of 0.15-0.21 g.

A flow injection method based on the catalytic action of iodide on the colour-fading reaction of the $FeSCN^{2+}$ complex was proposed and applied in order to determine iodine in milk. At pH 5.0, temperature 32°C and measurements at 460 nm, the decrease in absorbancy of Fe^{3+}-SCN (0.10 and 0.0020 mol /l) in the presence of NO_2^- (0.3 mol/ l) is proportional to the concentration of iodide, with a linear response up to 100.0 µg/l. The detection limit was determined as 0.99 µg/l and the system handles 48 samples per hour. Organic matter was destroyed by means of a dry procedure carried out under alkaline conditions. Alternatively, the use of a Schöninger combustion after the milk dehydration was evaluated. The residue was taken up in 0.12 mol/l KOH solubilization. For typical samples, recoveries varied from 94.5 to 105%, based on the amounts of both organic matter destroyed. The accuracy of the method was established by using a certified reference material (IAEA A-11, milk powder) and a manual method. The proposed flow injection method is now applied as an indicator of milk quality on the Brazilian market (de Araujo Nogueira et al., 1998).

Another spectrophotometric flow injection method for the determination of I- and based on the catalytic effect of this ion on the oxidation of pyrocatechol violet by potassium persulphate has been developed. The method allows the determination of 0.5–5 mg/l I- at a rate of 60 samples per hour and is subject to very little interference. It was successfully applied to the determination of iodide in table salt (Cerda et al., 1993).

4.1.4 Biological fluids and tissues

Due to the noninvasive way of sampling, urine is the most commonly analysed biological fluid. Efficient management of national salt iodization programmes depends on quality data on iodine concentrations in the urine and salt samples. These data are crucial in the evaluation of iodine interventions. Most of the analytical methods for urinary iodine concentration are based on the manual spectrophotometric measurement of Sandell-Kolthoff reduction reaction catalyzed by iodine using different oxidising reagents in the initial digestion step (Jooste & Strydom, 2010). Bilek et al. (Bilek et al., 2005) used a method which was based on alkaline ashing of urine specimens preceding the Sandell-Kolthoff reaction using brucine as a colorimetric marker. The detection limit was 2.6 µg/l and the limit of quantification was 11.7 µg/l, with intra-assay precision of 4% and inter-assay precision of 4.9%.

Another study described simple photometric determination of the iodine concentration in the thyroid tissue of small animals. Again, the method was based on the well-known catalytic Sandell-Kolthoff reaction. Prior to the analysis, the tissue was digested in a mixture of sodium chlorate and perchloric acid at 100 degrees C. Using this manner of digestion between 94 and 110% of iodine in the sample was recovered. Comparison with the neutron activation analysis showed excellent agreement of the obtained values (Tiran et al., 1991).

4.2 Chromatographic methods

While the main advantage of catalytic spectrophotometric methods is low cost of the needed equipment, chromatography is arguably the most widely used separation technique in the

A Review of Spectrophotometric and Chromatographic Methods and Sample Preparation Procedures for Determination
of Iodine in Miscellaneous Matrices

95

modern analytical laboratory. Fast, simple, reliable and sensitive chromatographic systems coupled with various detectors became the basic tool in many analytical laboratories. Routine analysis of iodine compounds can be carried out by means of gas chromatography (GC) and high performance liquid chromatography (HPLC). Analysis of inorganic iodine species in waters is mainly carried out with the use of ion chromatography (IC) or IC inductively coupled-mass spectrometry (IC-MS). Separation methods enable direct determination of various species of iodine in the presence of various kinds of complex components with the detection limit in the range of sub µg/l or µg/l (Hu et al., 1999; Schwer & Santschi, 2003). The IC method can separate I⁻ directly by using anion-exchange column, while HPLC method usually uses the reverse phase column modified by an ionpairing reagent in the mobile phase. Both spectrophotometric and electrochemical detectors are commonly used. Pulsed amperometric detection (PAD) typically utilizes gold, silver, platinum and glass carbon electrodes.

It has to be emphasized that both spectrophotometric and chromatographic methods are not applicable to a wide range of matrices. As far as complex matrices are concerned (e.g. seawater with high content of Cl⁻ and Br⁻ and relatively small of I⁻) IC and HPLC are useful tools for iodine determination. Unfortunately, none of these methods are flexible enough to measure all iodine species, including organo-iodine in biological samples. The oxidative pretreatment of biological materials limits the application of the described methods (IO_3^- and I⁻ can be converted to I_2 under acidic conditions). In order to analyse IO_3^-, I⁻, and organic forms of iodine in the same sample, IC is often coupled with ICP-MS. What is more, the coupling of highly efficient IC to multi-dimensional detectors such as MS or ICP/MS significantly increases sensitivity, while simultaneously reducing possible matrix interference to the absolute minimum.

4.2.1 Water samples

Liang et al. (Liang et al., 2005) applied the disposable electrode for the determination of iodide in soil and seawater samples with the spiked recovery ranging from 96–104% and the detection limit of 0.5 µg/L. Rong et al. (Rong et al., 2005) performed a direct determination of iodide and thiocyanate ions in seawater collected from the coasts of Japan. No sample pretreatment was needed. Liquid chromatography (LC) with a UV detection of 220 nm was applied. The separation was achieved on a C_{30} column of conventional size (150 mm × 4.6 mm i.d.) modified with poly(ethylene glycol). Such stationary phase enables the determination of I⁻ in seawater without any interference. Anions such as NO_3^-, NO_2^-, Br⁻ which absorb in the UV region do not interfere because the I⁻ peak is well resolved from the others. An aqueous solution of 300 mM sodium sulfate and 50 mM sodium chloride was used as the mobile phase. Detection limits (S/N=3) were obtained by injecting a 20-µL sample with 0.5 and 6 ng /ml for iodide and thiocyanate, respectively.

Buchberger (Buchberger, 1988) determined I⁻ (among other ions) in water samples using an anion-exchange stationary phase (Vydac 302-IC) and methanesulphonic acid solution as the mobile phase. A post-column reaction detector was developed based on the reaction between iodide or bromide, chloramine-T and 4,4' bis (dimethylamino)diphenylmethane. The detection limit was ca. 20 pg iodide injected.

A non-suppressed ion chromatography (IC) with inductively coupled plasma mass spectrometry (ICP-MS) was developed for simultaneous determination of trace IO_3^- and iodide in seawater. An anion-exchange column (G3154A/101, Agilent) was used for the separation of IO_3^- and I^- with an eluent containing 20 mM NH_4NO_3 at pH 5.6. NH_4NO_3 used in mobile phase minimizes salt deposition on the sampler and skimmer cones of mass spectrometer. Linear plots were obtained in a concentration range of 5.0–500 µg/l and the detection limit was 1.5 µg/L for IO_3^- and 2.0 µg/l for I^-. The proposed method was used to determine IO_3^- and I^- in seawaters without sample pre-treatment (with exception of dilution) (Chen et al., 2007).

Using IC-ICP-MS, Tagami and Uchida (Tagami & Uchida, 2006) measured concentrations of halogens (Cl, Br and I) in 30 Japanese rivers. Cesium was used as an internal standard during I counting. The typical detection limit was calculated as three times the standard deviation of the blank, between 0.01–0.04 µg/l. The ranges of geometric means of I in each river were 0.18-8.34 µg/l.

Bruggink et al. developed an anion-exchange chromatography method in combination with the pulsed amperometric detection (PAD) for the analysis of dissolved I^- in surface water and in absorption solutions obtained from adsorbable organic iodide (AOI) determination. The development of the amperometric waveform for a selective detection using a silver-working electrode together with the optimization of the injection volume and digital signal smoothing was performed. This method exhibited a detection limit of 0.02 µg/L, without any need of sample treatment other than micro-filtration. The results of AOI determination of the method described in this article were compared with results obtained with a different ion chromatography approach utilizing UV detection (Bruggink et al., 2007).

4.2.2 Seaweed

A gas chromatography (GC) method was reported for the trace analysis of I^- in processed seaweed by Lin et al. (Lin et al., 2003). The method is based on the derivatization of aqueous iodide extracted from seaweed with 2-(pentafluorophenoxy)ethyl 2 (piperidino) ethanesulfonate in toluene using tetra-n-hexylammonium bromide as a phase-transfer catalyst.

4.2.3 Food

GC method has been developed for determination of total iodine in food, based on the reaction of iodine with 3-pentanone. Organic matter of a sample was destroyed by an alkaline ashing technique. Iodide in a water extract of the ash residues was oxidized in order to free I_2 by adding $Cr_2O_7^{2-}$ in the presence of H_2SO_4. Liberated iodine reacted with 3-pentanone to form 2-iodo-3-pentanone, extracted into n-hexane, and then determined by gas chromatography with an electron-capture detector. Recoveries of I^- from spiked food samples ranged from 91.4 to 99.6%. Detection limit for iodine was 0.05 µg/g (Mitsuhashi & Kaneda Y, 1990).

Two methods were described for the preparation of samples for total iodine measurement in milk and oyster tissue. In the first method, the samples were combusted in a stream of oxygen to release iodine that, subsequently, was trapped in a solution as iodide. The second method used a new approach in which the samples were oxidized in a basic solution of peroxydisulfate. In this case, iodine was retained in the solution as an iodate. Total iodine was measured by means of the GC analysis of the 2-iodopentan-3-one derivative. The methods were tested using Standard Reference Materials (SRMs) 1549 Non-Fat Milk Powder, and 1566a and 1566 Oyster Tissue. Also, KI and KIO_3 were used for testing the procedures. The results obtained for the SRMs, given as average +/- standard deviation in µg/l, were: 3.39 +/- 0.14 and 3.40 +/- 0.23 for SRM 1549; 4.60 +/- 0.42 and 4.51 +/- 0.45 for SRM 1566a; and 2.84 +/- 0.16 and 2.76 +/- 0.06 for SRM 1566; values corresponding to combustion and wet oxidation, respectively. Overall, the absolute recoveries varied between 91 and 103% (Gu et al., 1997).

Cataldi and Ciriello (Cataldi & Ciriello, 2005) described a sensitive method based on anion-exchange chromatographic separation coupled with amperometric detection at a modified platinum electrode under constant applied potential (+0.85 V vs. Ag AgCl). An experimental setup with an in-line and very effective method of electrode modification was proposed using an amperometric thin-layer cross-flow detector and a flowing 300 mg/l solution of iodide. The working electrode was polarized to the limiting current for oxidation of iodide to iodine in acidic solutions with the consequent formation of an iodine-based film. The results confirmed that the modified electrode exhibits high analytical response for iodide electro-oxidation with a good stability and long-life. The detection limit of iodide was estimated to be 0.5 µg/l (S/N=3) with an injection volume of 50 µL. This method was applied successfully to quantify the iodide content of milk samples, wastewaters, common vegetables and solutions containing high chloride levels. The iodide peak was always observed without interferences from the excess of coexisting anions (e.g. Cl^-, SO_4^{2-} or Br^-). Chloride (the main component of marine samples) exhibited no effect upon the separation and detection of iodide. The same method (RP ion pair HPLC with an electrochemical detector and a silver working electrode) was considered by the International Organization for Standardization (the determination of iodide content of pasteurized whole milk and dried skimmed milk when present at levels from 0.03 µg/g to 1 µg/g and from 0.3 µg/g to 10 µg/g) (International Standard ISO, 2009).

Xu et al. (Xu et al., 2004) described a method for determination of iodate developed by RP-HPLC with UV detection. Iodate was converted to iodine, which was separated from the matrix using a reversed-phase Ultrasphere C18 column (250×4.6 mm, 5 µm) with methanol (1M) H_3PO_4 (1:4) as the mobile phase at 1.00 ml/min and UV detection at 224 nm. The calibration graph was linear from 0.05 µg/ml to 5.00 µg/ml for iodine with a correlation coefficient of 0.9994 (n=7). The detection limit was 0.01 µg/ml. The recovery was from 96% to 101% and the relative standard deviation was in the range of 1.5% to 2.9%.

A method based on the coupling of size-exclusion chromatography (SEC) with on-line selective detection of iodine by ICP MS was developed allowing determination of iodine species in milk and infant formulas. Iodine species were quantitatively eluted with 30 mM Tris buffer which was prepared by dissolving 30 mM of tris [tris(hydroxymethyl)-aminomethane] in water and adjusting the pH to 7.0 by the addition of hydrochloric acid (1 : 10, v/v) within 40 min and detected by ICP MS with a detection limit of 1 µg l[-1] (as I). A systematic study of iodine speciation in milk samples of different animals (cow, goat) and humans, of different geographic origin (several European countries) and in infant formulas from different manufacturers was carried out. When obtained after centrifugation of fresh milk or reconstituted , milk powders contained more than 95% of the iodine initially present in the milk of all the investigated samples with the exception of the infant formulas in which only 15-50% of the total iodine was found in the milk whey. Adding sodium dodecyl sulfonate (SDS) improved considerably the recovery of iodine from these samples (in case of the natural milk samples, this increase was ca. 10±20% but for infant formula samples the amount of iodine recovered in the supernatant was more than twice that in the samples not incubated with SDS). Iodine was found to be principally present as iodide in all the samples except infant formulas. In the latter, more than half of iodine was bound to a high molecular (>1000 kDa) species. The sum of all the species recovered from a size-exclusion column accounted for more than 95% of the iodine present in a milk sample. For the determination of total iodine in milk, a rapid method based on microwave-assisted digestion of milk with ammonia followed by ICP MS was optimized and validated using CRM 151 Skim Milk Powder (Fernandez-Sanchez & Szpunar, 1999).

4.2.4 Biological fluids and tissues

Odink et al. (Odink et al., 1988) presented a simple method for the routine analysis of iodide in urine. Iodide was separated by means of ion-pair reversed phase chromatography (RP-HPLC) and detected electrochemically with a silver electrode after a one-step sample clean-up. The coefficient of variation of a single analysis of iodide in a pooled urine sample (530 nmol/l) was 7.6%. The detection limit was 3 pmol (S/N 3), corresponding to 0.06 μmol/l. The recovery of iodide added to urine was 96±7 %.

There are also studies that compare spectrophotometric and RP-HPLC determinations of iodine concentrations in urine (Bier et al., 1998). In the first one ammonium persulfate was used as an oxidant in the modified ceric arsenite method. With the use of this sensitive method iodine concentrations can be determined in very small specimens (50 μL). A Technicon Autoanalyzer II and a paired-ion-RP HPLC were the basic analytical equipment. The authors found that the precision of this optimized ammonium persulfate method yielded inter assay CVs of <10% for urinary iodine concentrations >10 μg/dL. The detection limit was 0.0029 μg iodine. There was a high correlation between all three methods (r > 0.94 in any case) and the interpretation of the results was consistent. The authors suggested that the manual ammonium persulfate method could be performed in any routine clinical laboratory for urinary iodine analysis. Another benefit of the described methods is a possibility to process a large number of samples with high accuracy and minimal technician`s time.

When using the HPLC assay method, contaminations from the protein bound iodine do not interfere with the determination of the serum inorganic iodide (SII), making it the method of choice for detection in the serum. Although the clinical relevance of the measurement of SII is limited, it allows calculation of the absolute iodine uptake , which has a great value in certain pathophysiological studies (Rendl et al., 1998).

Błażewicz et al. (Błażewicz et al., 2011) examined correlations between the content of iodides in 66 nodular goitres and 100 healthy human thyroid tissues (50 - frozen and 50 formalin - fixed). A fast, accurate and precise ion chromatography method on the IonPac AS11 chromatographic column (Dionex, USA) with a pulsed amperometric detection (IC-PAD) followed by alkaline digestion with tetramethylammonium hydroxide (TMAH) in a closed system and with the assistance of microwaves was developed and used for the comparative analysis of two types of human thyroid samples (healthy and pathological). A good correspondence (for 10 additional determinations) between the certified (3.38 ± 0.02 ppm with variation coefficient /V.C. / of 0.59 % for Standard Reference Material (SRM) NIST 1549- non-fat milk powder) and the measured iodine concentrations (3.52 ± 0.29 ppm; V.C. = 10 %) was achieved. Suitability of the developed IC method was supported by validation results.

Ion chromatography coupled with electrospray ionization tandem mass spectrometry was applied for quantifying iodide, as well as perchlorate and other sodium-iodide symporter (NIS) inhibitors in the human amniotic fluid. The use of selective chromatography and tandem mass spectrometry decreased the need to clean up samples, leading to a quick and rugged method that is capable of the routine analysis of 75 samples per day. Along the physiologically relevant concentration range for the analytes, the analytical response was linear. The analysis of a set of 48 samples of amniotic fluid identified the range and median levels for iodide as: 1.7–170, 8.1µg/l (Blount & Valentin-Blasini, 2006).

5. Conclusion

There are many analytical methods available for detecting, and/or measuring iodine and its various species in complex matrices. Unfortunately, there is no perfect method which would be accurate, sensitive, cheap, fast, simple, and free of interferences at the same time. This review has been focused mainly on applications of spectophotometric and chromatographic methods of iodine analysis because they are widely used in practice, and relatively cheap. What is more, to achieve lower detection limits, they can also be coupled with other more sophisticated techniques (eg. ICP-MS). Although, these two methods have their own limitations, connected mainly with sample preatretment step (often timeconsuming), the literature data show continuous progress in the search for the best spectrophotometric and chromatographic conditions in iodine determinations. Reduction of time necessary for sample preparation still remains a challenge for analysts. Summarizing, future directions of iodine analysis lie rather in the simplification of methodologies and their extensive accessibility rather than in the tendency to decrease the limit of detection. Some recently published papers on the determination of iodine include: the evaluation of urinary iodide by the use of micro-photometric method compared to ICP-MS results (Grimm et al., 2011); determination of iodine and its species in plant samples using IC-ICP/MS (Lin et al., 2011); spectrophotometric determination of I^-, IO_3^-, IO_4^- in table salt, pharmaceutical preparations and sea water (George et al., 2011); investigation of the concentration-dependent mobility, retardation, and speciation of iodine in surface sediment from the river (Zhang et al., 2011); comparison of Sandell-Kolthoff reaction with potentiometric measurements of urinary iodide in female thyroid patients (Kandhro et al., 2011). One of the newest studies concerns the analysis of food samples by ICP-MS after alkaline digestion with TMAH (Tinggi et al., 2012). As it turns out, the newest published works utilize the most common already existing methods.

6. Acknowledgment

My appreciation and thanks are given to Prof. Ryszard Maciejewski, Vice Rector for Research at the Medical University of Lublin, for financial support of the research.

7. References

Adams, J.B., Holloway, C.E., George, F. & Quig, D. (2006). Analyses Of Toxic Metals And Essential Minerals In The Hair Of Arizona Children With Autism And Associated

Conditions And Their Mothers. *Biol Trace Elem Res*, Vol.110, pp 193–209, ISSN 1559-0720

Afkhami, A. & Zarei, A. R. (2001). Spectrophotometric Determination Of Periodate And Iodate By A Differential Kinetic Method. *Talanta* Vol.53, pp 815–820, ISSN *0039-9140*

Aiba, M., Ninomiya, J., Furuya, K., Arai, H., Ishikawa, H., Asaumi, S., Takagi, A., Ohwada, S. & Morishita, Y. (1999). Induction Of A Critical Elevation Of Povidone-Iodine Absorption In The Treatment Of A Burn Patient: Report Of A Case. *Surg Today* Vol.29, pp 157-159, ISSN 1436-2813

Allain, P., Berre, S., Krari, N., Laine-Cessac, P.,Le Bouil, A., Barbot, N., Rohmer,V.,Bigorgne, J. C. (1993). Use of plasma iodine assay for diagnosing thyroid Disorders. *J Clin Pathol.*, Vol.46, pp453-455, ISSN 0021-9746

Anderson, K. A. & Markowski, P. (2000). Speciation Of Iodide, Iodine, And Iodate In Environmental Matrixes By Inductively Coupled Plasma Atomic Emission Spectrometry Using In Situ Chemical Manipulation. *J. Aoac. Int.*, Vol.83, No.1, pp 225-230, ISSN 1944-7922

Andrási, E., Bélavári, C., Stibilj, V., Dermelj, M. & Gawlik, D. (2004). Iodine Concentration In Different Human Brain Parts. *Anal Bioanal Chem.*, Vol.378, No.1, pp 129-33, ISSN 1618-2642

Aumont, G. (1982). A Semi-Automated Method For The Determination Of Total Iodine In Milk. *Ann Rech Vet.* Vol.13, No.2, pp 205-210, ISSN 0003-4193

Balasubramanian, M. G. & Nagaraja K.S. (2008). Spectrophotometric Determination Of Iodine Species In Table Salt And Pharmaceutical Preparations. *Chem Pharm Bull (Tokyo)*, Vol.56, No.7, pp 888-893, ISSN 1347-5223

Bermejo-Barrera, P., Anllo-Sendín, R.Ma., Aboal-Somoza, M. & Bermejo-Barrera, A. (1999). Contribution To The Development Of Indirect Atomic Absorption Methods: Application Of The Ion Pair 1,10-Phenanthroline-Mercury(II)-Iodide To Iodide Determination In Water And Infant Formulae Samples. *Mikrochimica Acta*, Vol.131, No.3-4, pp 145-151, ISSN: 00263672

Bichsel, Y. & Von-Gunten, U. (1999). Determination of Iodide and Iodate by Ion Chromatography with Postcolumn Reaction and UV/Visible Detection. *Anal. Chem*, Vo.71, No.1, pp 34–38, ISSN 0003-2700

Bier, D., Rendl, J., Ziemann, M., Freystadt, D. & Reiners, Ch. (1998). Methodological And Analytical Aspects Of Simple Methods For Measuring Iodine In Urine. Comparison With HPLC And Technicon Autoanalyzer II. *Exp Clin Endocrinol Diabetes*, Vol.106, pp 27-31, ISSN 0947-7349

Bílek, R., Bednár, J. & Zamrazil, V. (2005). Spectrophotometric determination of urinary iodine by the Sandell-Kolthoff reaction subsequent to dry alkaline ashing. Results from the Czech Republic in the period 1994-2002. *Clin Chem Lab Med.* Vol.43, No.6, pp 573-580, ISNN 1434-6621

Błażewicz, A., Orlicz-Szczęsna, G., Szczęsny, P., Prystupa, A., Grzywa-Celińska & A., Trojnar, M. (2011). A Comparative Analytical Assessment Of Iodides In Healthy And Pathological Human Thyroids Based On IC-PAD Method Preceded By Microwave Digestion. *J Chromatogr B Analyt Technol Biomed Life Sci.*, Vol.15, No.879 (9-10), pp 573-578, ISSN1570-0232

A Review of Spectrophotometric and Chromatographic Methods and Sample Preparation Procedures for Determination
of Iodine in Miscellaneous Matrices

101

Blount, B. C. & Valentin-Blasini, L. (2006). Analysis Of Perchlorate, Thiocyanate, Nitrate And Iodide In Human Amniotic Fluid Using Ion Chromatography And Electrospray Tandem Mass Spectrometry. *Analytica Chimica Acta*, Vol.567, No.1, pp 87-93, ISSN 0003-2670

Braetter, P., Blasco, I.N., Negretti de Braetter, V. E. & Raab A. (1998). Speciation As An Analytical Aid In Trace Element Research In Infant Nutrition. *Analyst*, Vol.123, pp 821-826, ISSN 0003-2654

Bruggink, C., van Rossum, W. J. M., Spijkerman, E. & van Beelen, E. S. E. (2007). Iodide analysis by anion-exchange chromatography and pulsed amperometric detection in surface water and adsorbable organic iodide. *Journal of Chromatography A*, Vol. 1144, No.2, pp 170-174, ISSN 0021-9673

Buchberger, W. (1988). Determination Of Iodide And Bromide By Ion Chromatography With Post-Column Reaction Detection. *Journal of Chromatography A*, Vol.439, No.1, pp 129- 135, ISSN 0021-9673

Bürgi, H., Supersaxo, Z. & Selz, B.(1990). Iodine Deficiency Diseases In Switzerland One Hundred Years After Theodor Kocher's Survey: A Historical Review With Some New Goitre Prevalence Data. *Acta Endocrinol (Copenh)*, Vol.123, No.6, pp 577-590, ISSN 0001-5598

Cataldi, T. R. I., Rubino, A. & Ciriello, R. (2005). Sensitive Quantification Of Iodide By Ion-Exchange Chromatography With Electrochemical Detection At A Modified Platinum electrode. *Anal Bioanal Chem*, Vol.382, pp 134–141, ISSN 1618-2642

Cerda, A., Forteza, R. & Cerda, V.(1993). Determination Of Iodide In Table Salt By Flow Injection Analysis Using Pyrocatechol Violet. *Food Chemistry*, Vol.46, No.1, pp 95 - 99, ISSN 0308-8146

Chen, Z. L., Megharaj, M. & Naidu, R. (2007). Speciation Of Iodate And Iodide In Seawater By Non-Suppressed Ion Chromatography With Inductively Coupled Plasma Mass Spectrometry. *Talanta*, Vol.72, No.5, pp 1842-1846, ISSN 0039-9140

Cicerone, R. J. (1981). Halogens in the Atmosphere. *Rev. Geophys. Space Phys.*, Vol.19, pp 123-139, ISSN 0034-6853

Das, P., Gupta, M., Jain, A. & Verma, K. K. (2004). Single Drop Microextraction Or Solid Phase Microextraction-Gas Chromatography-Mass Spectrometry For The Determination Of Iodine In Pharmaceuticals, Iodized Salt, Milk Powder And Vegetables Involving Conversion Into 4-Iodo-N,N-Dimethylaniline. *J. Chromatogr. A*, Vol.1023, pp 33–39, ISSN: 00219673

De Araujo Nogueira, A. R., Mockiuti, F., De Souza, G. B. & Primavesi, O. (1998). Flow Injection Spectrophotometric Catalytic Determination of Iodine in Milk. *Anal. Sci.*, Vol.14, pp 559-564, ISSN 1348-2246

Delange, F., de Benoist, B., Burgi, H. (2002). Determining median urinary iodine concentration that indicates adequate iodine intake at population level. *Bull World Health Organ.*, Vol.80, No.8, pp633-666, ISSN 0042-9686

Ensafi, A. & Dehaghi, G.B. (2000). Flow-Injection Simultaneous Determination Of Iodate And Periodate By Spectrophotometric And Spectrofluorometric Detection. *Anal. Sci.* Vol.16, pp 61–64, ISSN 09106340

Fang, M.C., Stafford, R.S., Ruskin, J.N., & Singer, D.E. (2004). National Trends In Antiarrhythmic And Antithrombotic Medication Use In Atrial Fibrillation. *Arch Intern Med*, Vol.164, pp 55-60, ISSN 1538-3679

Fecher P.A. & A. Nagengast. (1994). Trace analysis in high matrix aqueous-solutions using helium microwave-induced plasma-mass spectrometry. *J. Anal. At. Spectrom.*, Vol.9, pp1021-1027, ISSN 1364-5544

Fecher P.A., Goldman, I. & A. Nagengast. (1998). Determination Of Iodine In Food Samples By Inductively Coupled Plasma Mass Spectrometry After Alkaline Extraction. *Journal of Analytical Atomic Spectometry*, Vol.13, pp 977-982, ISSN 1364-5544

Fernández-Sánchez, L. & Szpunar, J. (1999). Speciation Analysis For Iodine In Milk By Size - Exclusion Chromatography With Inductively Coupled Plasma Mass Spectrometric Detection (SEC-ICP MS). *J.Anal. At. Spectrom.*, Vol.14, pp 1697-1702, ISSN 1364-5544

Fernández-Sánchez, L., Bermejo-Barreraa, P., Fraga-Bermudez, Szpunar, J. & Łobiński R. (2007). Determination Of Iodine In Human Milk And Infant Formulas. *Journal of Trace Elements in Medicine and Biology*, Vol.21, Supp.1, pp 10-13, ISSN 0946-672X

Fischer, P.W.F., Labbe, M. R. & Giroux, A. (1986). Colorimetric determination of total iodine in foods by iodide-catalyzed reduction of Ce+4. *J. Assoc. Off. Anal. Chem.* , Vol.69, No.4, pp 687 – 689, ISSN 0004-5756

FOOD STANDARDS AGENCY (FSA) UK – 2000. Total Diet Study - Fluorine, Bromine and Iodine, Available from http://www.food.gov.uk/science/surveillance/fsis2000/5tds

Fragu, P., Briançon, C., Fourré, C., Clerc, J., Casiraghi, O., Jeusset, J., Omri, F. & Halpern, S. (1992). SIMS Microscopy In The Biomedical Field. *Biol Cell.*, Vol.74, No.1, pp 5-18, ISSN 1768-322X

Fuge, R., Johnson, C. C. (1986).The geochemistry of iodine - a review. *Environmental Geochemistry and Health,*Vol.8, No. 2, pp 31-54, ISSN 0269-4042

Gamallo-Lorenzo,D., Barciela-Alonso, M. C., Moreda-Pineiro, A., Bermejo-Barrera, A. & Bermejo-Barrera, P. (2005). Microwave-assisted alkaline digestion combined with microwave-assisted distillation for the determination of iodide and total iodine in edible seaweed by catalytic spectrophotometry. *Analitica Chimica Acta, Vol.* 542, pp 287-295, ISSN 0003-2670

Garland, J. A. & Curtis, H. J. (1981). Emission of iodine from the sea surface in the presence of ozone. *J. Geophys. Res.*, Vol.86 (C4), pp 3183–3186, ISSN 0148-0227

George, M., Nagaraja, K. S., Natesan Balasubramanian, N. (2011). Spectrophotometric Determination of Iodine Species in Table Salt,Pharmaceutical Preparations and Sea Water., *Eurasian J Anal Chem.*, Vol.6, No.2, pp129-139, ISSN 1306-3057

Gilfedder, B.S., Althoff, F., Petri, M. & Biester, H. (2007). A Thermo Extraction-UV/Vis Spectrophotometric Method For Total Iodine Quantification In Soils And Sediments. *Analytical And Bioanalytical Chemistry*, Vol.389, No(7-8), pp. 2323-2329, ISSN 1618-2642

Gottardi, W. (1998). Redox-potentiometric/titrimetric analysis of aqueous iodine solutions. *Fresenius J Anal Chem*, Vol.362, pp263–269, ISSN 0937-0633

Grimm, G., Lindorfer, H., Kieweg, H., Marculescu, R., Hoffmann, M., Gessl, A., Sager, M.,Bieglmayer, C. (2011). A simple micro-photometric method for urinary iodine determination. *Clin Chem Lab Med.*, Vol.49, pp1749-51, ISSN 0008-4212

Gu, F., Marchetti, A.A. & Straume, T. (1997). Determination Of Iodine In Milk And Oyster Tissue Samples Using Combustion And Peroxydisulfate Oxidation. *Analyst.* Vol.122, No.6, pp 535-537, ISSN 0003-2654

Guenther, E. A., Johnson, K. S. & Coale, K. H. (2001). Direct Ultraviolet Spectrophotometric Determination of Total Sulfide and Iodide in Natural Waters. *Anal. Chem.* Vol.73, | pp 3481-3487, ISSN 0003-2700

Gurkan, R., Bicer, N., Ozkan, M. H., & Akcay, M. (2004). Determination Of Trace Amounts Of Iodide By An Inhibition Kinetic Spectrophotometric Method. *Turk J Chem,* Vol.28, pp 181-191, ISSN 1303-6130

Hansson, M., Isaksson, M. & Berg, G. (2008). Sample Preparation For In Vitro Analysis Of Iodine In Thyroid Tissue Using X-Ray Fluorescence. *Cancer Inform.* Vol.6, pp 51-57, ISSN 1176-9351

Heckwan, M.M. (1979). Analysis of foods for iodine: interlaboratory study. *J Assoc Off Anal Chem,* Vol.62, pp 1045- 1049, ISSN 0004-5756

Hemken, R.W. (1979). Factors That Influence The Iodine Content Of Milk And Meat: A Review. *J Anim Sci.* Vol.48, pp 981-985, ISSN1525-3163

Hou, X. (2009). Iodine Speciation in Foodstuffs, Tissues, and Environmental Samples, In: Comprehensive Handbook of Iodine, Preedy, (Ed.), 139-150, Elsevier, ISBN 978- 0-12-374135-6, Burlington, USA

Hou, X., Chai, C. & Qian, Q. (1997). Determination Of Bromine And Iodine In Biological And Environmental Materials Using Epithermal Neutron Activation Analysis. Fresenius Anal Chem, Vol.357, pp 1106–1110, ISSN 1432-1130

Hou, X., Dahlgaard, H., Rietz, B., Jacobsen U., Nielsen, S. P. & Aarkrog, A. (1999). Determination of Chemical Species of Iodine in Seawater by Radiochemical Neutron Activation Analysis Combined with Ion-Exchange Preseparation. Anal. Chem., Vol.71, No. 14, pp 2745–2750, ISSN 0003-2700

Hu, W., Haddad, P. R., Hasebe, K., Tanaka, K., Tong, P. & Khoo, C. (1999). Direct Determination of Bromide, Nitrate, and Iodide in Saline Matrixes Using Electrostatic Ion Chromatography with an Electrolyte as Eluent. *Anal. Chem.,* Vol.71, pp 1617-1620, ISSN 0003-2700

International Standard ISO 14378, IDF 167, 2009 -10-01. Milk And Dried Milk - Determination Of Iodide Content - Method Using High-Performance Liquid Chromatography, Available from http://www.iso.org/iso/iso_catalogue/catalogue_tc/catalogue_detail.htm?csnumber=53832

Johnson, C. C. (2003).The geochemistry of iodine and its application to environmental strategies for reducing the risks from iodine deficiency disorders. *British Geological Survey Commissioned Report,* CR/03/057N. pp 54

Jooste, P.L. & Strydom, E. (2010). Methods For Determination Of Iodine In Urine And Salt. *Best Pract Res Clin Endocrinol Metab.,* Vol.24, No.1, pp 77-88, ISSN 1521-690X

Jopke, P., Bahadir, M., Fleckenstein, J. & Schnug, E. (1996). Iodine Determination In Plant Materials. *Communications In Soil Science And Plant Analysis,* Vol.27, No.3-4, pp 741-751, ISSN 1532-2416

Jungreis, E. & Gedalia, I. (1960). Ultramicro Determination Of Iodine In Drinking Water On The Basis Of Feigl's Catalytic Reaction. *Microchimica Acta,* Vol.48, No.1, pp 145-149, ISSN 1436-5073

Kamavisdar, A. & Patel, R. M. (2002). Flow Injection Spectrophotometric Determination of Iodide in Environmental Samples, *Microchimica Acta,* Vol.140, No.1-2; pp. 119-124, ISSN 1436-5073

Kandhro, G. A., Kazi, T. G., Kazi, S. N., Afridi, H. I., Arain, M. B., Baig, J. A., Shah, A. Q. & Syed, N. (2009). Evaluation of the Iodine Concentration in Serum and Urine of Hypothyroid Males Using an Inexpensive and Rapid Method. *Pak. J. Anal. Environ. Chem.*, Vol. 10, No. 1 & 2, pp 67-75, ISSN 1996-918X

Kandhro, G. A., Gul Kazi,T. G., Sirajuddin, Kazi, N., Afridi, H. I., Arainc, M. B., Baig, J. A., Shah, A. Q., Wadhwa, S. K., Shah, F. (2011). Comparison of Urinary Iodide Determination in Female Thyroid Patients by two Techniques. *Russian Journal of Electrochemistry.*, Vol.47, No.12, pp 1355-1362, ISSN 1023-1935

Kesari, R., Rastogi, R. & Gupta, V. K. (1998). A Simple And Sensitive Spectrophotometric Method For The Determination Of Iodine In Environmental Samples. *Chemia Analityczna* , Vol.43, No.2, pp 201-207, ISSN 0009-2223

Knapp, G. & H. Spitzy, H. (1969). Untersuchungen Zur Optimierung Der Reaktionsbedingungen Für Die Katalytische Jodwirkung Auf Das System CE(IV)-Arsenige Säure (Eine Modifizierte Sandell-Kolthoffreaktion). *Talanta,* Vol.16, No.10, pp 1353-1360, ISSN 0039-9140

Knapp, G., Maichin, B., Fecher, P., Hasse, S. & Schramel, P. (1998). Iodine Determination In Biological Materials Options For Sample Preparation And Final Determination. *Fresenius J Anal Chem,* Vol. 362, pp 508 – 513, ISSN 1432-1130

Koh, T., Ono, M. & Makino, I. (1988). Spectrophotometric determination of iodide at the 10–6 mol l-1 level by solvent extraction with methylene blue. *Analyst,* Vol.113, pp 945-948, ISSN 0003-2654

Konieczka, P. & Namieśnik J. (Ed.), Ocena i kontrola jakości wyników pomiarów analitycznych. Wydawnictwa Naukowo-Techniczne, ISBN 978-83-204-3255-8, Warszawa, 2007

Levine, K.E., Essader, A.S., Weber, F.X., Perlmutter, J.M., Milstein, L.S., Fernando, R.A., Levine, M.A., Collins, B.J., Adams, J.B. & Grohse, P.M. (2007).Determination Of Iodine In Low Mass Human Hair Samples By Inductively Coupled Plasma Mass Spectrometry.*Bull Environ Contam Toxicol,* Vol.79, No.4, pp 401-404, ISSN 0007-4861

Liang, L., Cai, Y., Mou, S. & Cheng, J. (2005). Comparisons Of Disposable And Conventional Silver Working Electrode For The Determination Of Iodide Using High Performance Anion-Exchange Chromatography With Pulsed Amperometric Detection. *Journal of Chromatography A,* Vol.1085, No.1, pp 37-41, ISSN 0021-9673

Lin, F.-M., Wu, H.-L., Kou, H.-S. & Lin, S.-J. (2003). Highly Sensitive Analysis of Iodide Anion in Seaweed as Pentafluorophenoxyethyl Derivative by Capillary Gas Chromatography. J. Agric. Food Chem., Vol.51, No.4, pp 867–870 , ISSN 0021-8561

Lin, L., Chen, G., Chen, Y. (2011). Determination of iodine and its species in plant samplesusing ion chromatography-inductively coupled plasma mass spectrometry. *Chinese Journal of Chromatography,* Vol.29, pp662-666, ISSN 1872-2059

Lovelock, J.E., Maggs, R.J. & Wade, R.J. (1973). Halogenated Hydrocarbons In And Over The Atlantic. *Nature,* Vol.241, No. 5386, pp 194-196, ISSN 0028-0836

Lu, Y.-L., Wang, N.-J., Zhu, L., Wang, G.-X., Wu, H., Kuang, L. & Zhu W.-M. (2005). Investigation Of Iodine Concentration In Salt, Water &Soil Along The Coast Of Zhejiang, China. *J Zhejiang Univ Sci B.*, Vol.6, No.12, pp 1200–1205, ISSN 1862-1783

Mani, D., Gnaneshwar Rao T., Balaram, V., Dayal, A.M. & Kumar, B. (2007). Rapid Determination Of Iodine In Soils Using Inductively Coupled Plasma Mass Spectrometry. *Current Science,* Vol.93, No. 9, 10, pp 1219-1221, ISSN 0011-3891

Maros, L., Kaldy, M. & Igaz, S. (1989). Simultaneous Determination Of Bromide And Iodide
     As Acetone Derivatives By Gas Chromatography And Electron Capture Detection
     In Natural Waters And Biological Fluids. *Anal. Chem.*, Vol. 61, 733–735, ISSN 0003-
     2700
Michalke, B. (2006). Trace Element Speciation In Human Milk. *Pure Appl. Chem.*, Vol.78,
     No.1, pp. 79-90, ISSN 0033-4545
Mitsuhashi, T. & Kaneda, Y. (1990). Gas Chromatographic Determination Of Total Iodine In
     Foods. *J Assoc Off Anal Chem.*, Vol.73, No.5, pp 790-792, ISSN 0004-5756
Miyake, Y. & Tsunogai, S. (1963). Evaporation of iodine from the ocean. *J. Geophys. Res.*,
     Vol.68, pp 3989-3993, ISSN 0148-0227
Moxon, R.E. & Dixon, E.J. (1980). Semi-Automatic Method For The Determination Of Total
     Iodine In Food. *Analyst*, Vol.105, pp 344–352, ISSN 0003-2654
Mura, P., Papet, Y., Sanchez, A. & Piriou, A. (1995). Rapid And Specific High-Performance
     Liquid Chromatographic Method For The Determination Of Iodide In Urine, *J.
     Chromatogr. B*, Vol.664, pp 440–443, ISSN 1570-0232
Muramatsu, Y., Uchida, S., Fehn, U., Amachi, S. & Ohmomo, Y. (2004). Studies With
     Natural And Anthropogenic Iodine Isotopes: Iodine Distribution And Cycling In
     The Global Environment. *J. Environ. Radioact.*, Vol.74, pp 221–223, ISSN 0265-931X
Naozuka J., Mesquita Silva da Veiga M.A., Oliveira P.V. & de Oliveira E. (2003).
     Determination Of Chlorine, Bromine And Iodine In Milk Samples By ICP-OES. *J.
     Anal. Atomic Spectrom.* Vol.18, pp 917–921, ISSN 1364-5544
Niedobová, E., Machát, J., Otruba, V. & Kanicky, V. (2005). Vapour Generation Inductively
     Coupled Plasma Optical Emission Spectrometry In Determination Of Total Iodine
     In Milk. *J.Anal.Atom. Spectrom.*, Vol.20, pp 945–949, ISSN 1364-5544
Odink, J., Bogaards, J.J.P., Sandman, H., Egger, R.J., Arkesteyn, G.A. & de Jong, P. (1988).
     Excretion Of Iodide In 24-H Urine As Determined By Ion-Pair Reversed-Phase
     Liquid Chromatography With Electrochemical Detection. *Journal Of Chromatography
     B: Biomedical Sciences And Applications* Vol.431, pp 309-316, ISSN 1570-0232
Okerlund, M.D. (1997). The Clinical Utility of Fluorescent Scanning of the Thyroid. In:
     Medical Applications of Fluorescent Excitation Analysis, Kaufman & Price (Ed.),
     149-160, CRC Press, ISBN 0849355079, Boca Raton Florida, USA
Pearce, E.N., Gerber, A.R., Gootnick, D.B., Khan, L., Li, R., Pino, S. & Braverman, L. E.(2002).
     Effects Of Chronic Iodine Excess In A Cohort Of Long-Term American Workers In
     West Africa. *J ClinEndocrinol Metab*, Vol.87, pp 5499 –5502, ISSN 1945-7197
Pennington, J.A.T., Schoen, S.A., Salmon, G.D., Young, B., Johnson, R.D. & Marts, R.W.
     (1995). Composition of Core Foods of the U.S. Food Supply, 1982-1991. III. Copper,
     Manganese, Selenium, and Iodine. *J Food Comp Anal.*, Vol.8, No.2, pp 171-217, ISSN
     1096-0481
Rasmussen, R. A., Khalil, M.A.K., Gunawardena, R. & Hoyt, S. D. (1982). Atmospheric
     Methyl Iodide (CH_3I). *J. Geophys. Res.*, Vol.87, pp 3086-3090, ISSN 0148-0227
Rendl, J., Bier, D. & Reiners, C. (1998). Methods For Measuring Iodine In Urine And Serum.
     *Exp Clin Endocrinol Diabetes,*Vol.106, Suppl 4, pp 34-41, ISSN 09477349
Rodriguez, P. A. & Pardue, H. L. (1969). Kinetics of the Iodide-Catalyzed Reaction between
     Cerium(IV) and Arsenic(III) in Sulfuric Acid Medium. *Anal. Chem.*, Vol.41, pp
     1369-1376, ISSN 0003-2700

Rognoni, J. & Simon, C. (1974). Critical Analysis Of The Glutaraldehyde Fixation Of The Thyroid Gland: A Double-Labelling Experiment. *J. Microscopie., Vol.* 21, pp 119–128, ISSN 1365-2818

Rong, L., Lim, L. W. & Takeuchi, T. (2005). Determination of Iodide and Thiocyanate in Seawater by Liquid Chromatography with Poly(ethylene glycol) Stationary Phase. *Chromatographia,* Vol.61, pp 371-374, ISSN 1612-1112

Sandell, E. B. & Kolthoff, I. M. (1934). Chronometric catalytic method for the determination of micro quantities of iodine. *J. Am. Soc.*, 56, pp 1426-1435 ISSN: 0002-7863

Sandell, E. B. & Kolthoff, I. M. (1937). Micro Determination of Iodine by a Catalytic Method. *Microchim. Acta*, Vol.1, pp 9-25, ISSN 00263672

Schramel, P. & Hasse, S. (1994). Iodine Determination In Biological Materials By ICP-MS. *Microchimica Acta,* Vol.116, No.4, pp 205-209, ISSN 1436-5073

Schwehr, K. A. & Santschi, P. H. (2003). Sensitive Determination Of Iodine Species, Including Organo-Iodine, For Freshwater And Seawater Samples Using High Performance Liquid Chromatography And Spectrophotometric Detection. *Anal. Chim. Acta*, Vol.482, pp 59-71, ISSN 00032670

Sheppard, M.I., Thibault, D.H., Mcmurray, J. & Smith, P. A. (1995). Factors Affecting The Soil Sorption Of Iodine. *Water Air Soil Pollut.*, Vol.83, No.1-2, pp 51-67, ISSN 0049-6979

Silva, R. L., De Oliveira, A. F. & Neves, E. A. (1998). Spectrophotometric Determination Of Iodate In Table Salt. *J. Braz. Chem. Soc.,*Vol.9, No.2, pp 171-174, ISSN 0103-5053

Simon, R., Tietge, J. E., Michalke, B. Degitz, S. & Schramm K.-W. (2002). Iodine Species And The Endocrine System: Thyroid Hormone Levels In Adult Danio Rerio And Developing Xenopus Laevis. *Anal Bioanal Chem, Vol.* 372, pp 481–448, ISSN 1618-2642

Stolc, V. (1961). Interference of Certain Ions with the Catalytic Action of Iodine in the Sandell-Kolthoff Reaction *Fresenius J Anal Chem*, Vol.183, No.4, pp262–267, ISSN 0937-0633

Szybiński, Z. (2009). Iodine prophylaxis in Poland in the lights of the WHO recommendation on reduction of the daily salt intake. *Pediatric Endocrinology, Diabetology and Metabolism.*, Vol.15, No.2, pp103-107, ISSN 2083-8441

Tadros, T.G., Maisey, M.N., Ng Tang Fui S.C. & Turner, P,C.(1981). The Iodine Concentration In Benign And Malignant Thyroid Nodules Measured By X-Ray Fluorescence, *Br J Radiol.*, Vol.54, No.643, pp 626-629, ISSN 1748-880X

Tagami, K. & Uchida, S. (2006). Concentrations Of Chlorine, Bromine And Iodine In Japanese Rivers. *Chemosphere*, Vol.65, No.11, pp 2358-2365, ISSN 0045-6535

Teas, J., Pino, S., Critchley, A. & Braverman, L.E. (2004). Variability Of Iodine Content In Common Commercially Available Edible Seaweeds. *Thyroid, Vol.*14, No.10, pp 836-841, ISSN 1050-7256

Teng, W.; Shan, Z. & Teng, X. (2009). Effect of an Increased Iodine Intake on Thyroid, In: Comprehensive Handbook of Iodine, Preedy, (Ed.), 1213-1220, Elsevier, ISBN 978-0-12-374135-6, Burlington, USA

Tinggi, U., Schoendorfer, N. Davies, P. S. W., Scheelings, P., Olszowy, H. (2012). Determination of iodine in selected foods and diets by inductively coupled plasma-mass spectrometry., *Pure Appl. Chem.*, Vol.84, No.2, pp 291-299, ISSN1365-3075

Tiran, B., Wawschinek, O., Eber, O., Beham, A., Lax, S. & Dermelj, M. (1991). Simple Determination Of Iodine In Small Specimens Of Thyroid Tissue. *Exp Clin Endocrinol.*, Vol.98, No.1, pp 32-36, ISSN 0232-7384

Trokhimenko, O.M. & Zaitsev, V.N. (2004). Kinetic Determination of Iodide by the Sandell-Kolthoff Reaction Using Diphenylamine-4-Sulfonic Acid. *J Anal Chem.*, Vol.59,pp 491–494, ISSN 1608-3199

United Nations Administrative Committee on Coordination/ Sub-Committee on Nutrition (2000) 4th Report on World Nutrition Situation – Nutrition Throughout the Life Cycle. Available from
http://www.unsystem.org/scn/archives/rwns04/index.htm (accessed June 2007)

U.S. Food and Drug Administration, Code of Federal Regulations, CFR 21, Sections 184.1634 and 184.1265. Revised April 1, 2009. Available from
http://ods.od.nih.gov/factsheets/Iodine-HealthProfessional

U.S. Salt Institute, accessed July 9, 2007, Available from http://www.saltinstitute.org/html

Varga, I. (2007). Iodine Determination In Dietary Supplement Products By TXRF And ICP-AES Spectrometry. *Microchemical Journal*, Vol.85, pp 127-131, ISSN 0026265X

Verheesen, R. H. & Schweitzer, C. M. (2011). Iodine and Brain Metabolism. In: Handbook of Behavior, Food and Nuirtion, Preedy et al., (Ed.), 2411-2425, Springer, ISBN 978- 0-387-92270-6, New York Dordrecht Heidelberg London

Whitehead, D.C. (1984). The Distribution And Transformations Of Iodine In The Environment. Environ. Int., Vol.10, pp 321–339, ISSN 0160-4120

World Health Organization. United Nations Children's Fund & International Council for the Control of Iodine Deficiency Disorders. (2007). Assessment Of Iodine Deficiency Disorders And Monitoring Their Elimination. In: A Guide For Programme Managers, 3rd Ed. Geneva, Switzerland, pp 1-108, Available from

Xu, X. R., Li, H. B., Gu, J. D. & Paeng, K. J. (2004). Determination of Iodate in Iodized Salt by Reversed-Phase High-Performance Liquid Chromatography with UV Detection. *Chromatographia*, Vol.60, No.11-12, pp 721-723, ISSN 1612-1112

Yebra, M.C. & Bollaín, M.H. (2010). A Simple Indirect Automatic Method To Determine Total Iodine In Milk Products By Flame Atomic Absorption Spectrometry. *Talanta*, Vol.82, No.2, pp 828-33, ISSN 0039-9140

Yoshida, S. & Muramatsu, Y. (1995). Determination Of Organic, Inorganic And Particulate Iodine In The Coastal Atmosphere Of Japan. *Journal Of Radioanalytical And Nuclear Chemistry*, Vol.196, No. 2, pp 295-302, ISSN 1588-2780

Zabala, J., Carrión, N., Murillo, M., Quintana, M., Chirinos, J., Seijas, N., Duarte, L. & Brätter, P. (2009). Determination Of Normal Human Intrathyroidal Iodine In Caracas Population. *Trace Elem Med Biol.*, Vol.23, No.1, pp 9-14, ISNN 0946-672X

Zaichick, V. & Zaichick, S. (1997). Normal Human Intrathyroidal Iodine. *Sci Total Environ.* Vol.27, No.206(1), pp 39-56, ISSN 0048-9697

Zhang, S., Du, J., Xu, C., Schwehr, K. A., Ho, Y.-F., Li, H.-P., Roberts, K. A., Kaplan, D. I., Brinkmeyer, R., Yeager, C. M., Chang, Hyun-shik, P. H. Santschi, P. H. (2011). Concentration-Dependent Mobility, Retardation, and Speciation of Iodine in Surface Sediment from the Savannah River Site. Environ. Sci. Technol., Vol.45, No.13, pp 5543–5549, ISSN 0013-936X

Zava, T. T. & Zava, D. T. (2011). Assessment of Japanese iodine intake based on seaweed consumption in Japan: A literature-based analysis. *Thyroid Research*, Vol.4, p 14, ISSN 1756-6614

Zhai, Q.-Z., Zhang, X.-X. & Goupages, X.-L. (2010). Catalytic Kinetic Spectrophotometric Determination Of Trace Amounts Of Iodine. *Instrumentation Science & Technology*, Vol.38, No.2, pp 125-134, ISSN 1073-9149

Zimmermann, M. (2009). Iodine Deficiency. *Endocr Rev.*, Vol.30, No.4, pp 376-408, ISSN 0163-769X

6

Optical and Resonant Non-Linear Optical Properties of J-Aggregates of Pseudoisocyanine Derivatives in Thin Solid Films

Vladimir V. Shelkovnikov[1] and Alexander I. Plekhanov[2]
[1]Novosibirsk Institute of Organic Chemistry SB RAS, Novosibirsk,
[2]Institute of Automation and Electrometry SB RAS, Novosibirsk,
Russia

1. Introduction

The properties of colligative states of spontaneously aggregated polymethine dyes differ substantially from those of monomeric dye. Excellent examples of such self-organized molecular ensembles are J-aggregates of cyanine dyes. The J-aggregated state is now being considered for a number of non-cyanine dyes, the cyanine dye is still the most known and effective dye for J-aggregate formation (Wurthner, 2011). J-aggregates of cyanine dyes, first discovered by Jelley and Scheibe in 1936 (Jelley, 1936; Scheibe, 1936), have been studied for many years (Kobayashi, 1996). J-aggregates of cyanine dyes attract the attention of the researchers due to their interesting optical properties. J-aggregates are characterized by a strong absorption peak (J-peak) with narrow line widths which are bathochromically shifted relative to the absorption band of the monomeric dye. Their role as photographic sensitizers can hardly be overestimated (Tani, 1996; Trosken et al., 1995; Shapiro, 1994). Aggregates of dye molecules may be used to mimic light harvesting arrays and to prepare artificial photosynthetic systems (McDermott et al., 1995; Blankenship, 1995). Another development is the efficient electroluminescence revealed in single-layer light-emitting diodes based on electron-hole conducting polymers containing nano-crystalline phases of J-aggregates of cyanine dyes (Mal'tsev et al., 1999).

The promise in the property of J-aggregates lies in their high non-linear cubic optical susceptibility, $\chi^{(3)} \sim 10^{-7}$ esu, with a fast response time at the J-peak resonance in solutions and polymer films (Wang, 1991; Bogdanov et al., 1991).

The pseudoisocyanine dye (PIC) is the known dye which forms the J-aggregates in solutions. Of particular interest is the formation and non-linear optical properties of J-aggregates in thin solid films. Films of J-aggregates of organic dyes are promising nanomaterials for non-linear optical switches because they have the unique properties of high non-linear bleaching and non-linear refraction (Markov et al., 2000). As shown in Glaeske et al. (2001), films of J-aggregates with bistable behaviour may be the basis for two-dimensional optical switches, controllable by light. The non-linear optical properties of organic dye J-aggregates have been intensively studied for application in future optical telecommunication and signal processing systems with ultrahigh bit rates (Tbit/s) (Furuki et al., 2000). The observed giant resonant third-order susceptibility in PIC thin solid films $\chi^{[3]}\sim10^{-5}$-10^{-4} esu (5 orders of magnitude greater than in polyconjugated polymers) and accessible production of optical quality films over a large area gives the possibility of J-aggregates application in telecommunication for terahertz demultiplexing of optical signals.

The methods for obtaining the J-aggregates in solutions do not give stable aggregates, which hampers their application as non-linear optical materials. Besides, a thin film geometry is preferable for applications. Therefore, the preparation of large area thin films of J-aggregates with thermal and photochemical stability, and high optical quality is vital for practical application and a matter of much current interest. The pseudoisocyanine at proper conditions efficiently forms J-aggregates in solid thin films (Shelkovnikov et al., 2002). The aim of this paper is to clarify the influence on the spectral linear and non-linear properties of PIC J-aggregates in thin solid films using a number of factors: structure of pseudoisocyanine dye derivatives, local field factor and character of J-peak broadening.

2. Experiments and discussions

2.1 Experimental part - Materials and methods

We used two methods to stimulate the J-aggregates' formation in the thin solid films. The use of a solution of PIC dyes with long alkyl substituents (Shelkovnikov et al., 2002) and the use of a PIC2-2 solution with the addition of cluster hydroborate anions (Shelkovnikov et al., 2004). Synthesis of the number of the derivatives of PIC with symmetrical and non-symmetrical long alkyl chain substituters for the physical-chemical experiments was carried out and the obtained dyes were isolated, chromatographically refined and characterized using the ^{1}H nuclear magnetic resonance method (Orlova et al., 2002; Orlova et al., 1995). Synthesis and characterization of the PIC2-2 derivative with cluster anion - closo-hexahydrodecaborate ($B_{10}H_{10}^{2-}$) by the infra-red spectroscopy and X-ray spectroscopy methods was carried out [Plekhanov et al., 1998a, 1998c; Cerasimova et al., 2000).

The thin solid films of PIC dye were prepared using spin-coating of the dye solution on the clean glass plates 2,5x2,5 cm^2 with a rate of rotation of 2000-3000 rpm on the custom made spin-coating equipment.

The linear optical spectra of the thin solid samples, depending on the experiments, were measured on the spectrophotometer Hewlett Packard 8453, fast fibre optics spectrophotometer Avantes AVS-SD2000 and fast fibre optics spectrophotometer Calibri VMK Optoelectronics, Novosibirsk (http://www.vmk.ru/). Fast spectrophotometers allow the measurement of 40 spectra per second. Sample heating for determination of the J-aggregates' thermal stability was done in a thermostatic optical chamber with a constant heating rate of 1.5 deg/min.

The thickness, dispersion of refractive index, and absorption of the obtained films were measured on a spectral ellipsometer Ellips developed at the A.V.Rzhanov Institute of Semiconductor Physics SB RAS (http://www.isp.nsc.ru/). The optical parameters of the film according to ellipsometric parameters delta (Δ) and psi (Ψ) were found by approximating the single-phase model of the Si-substrate/absorbing film.

The steady-state luminescence spectra were recorded on the Cary-Eclipse (Varian) and Hitachi 850 spectrofluorimeters. The kinetics of the luminescence decay was measured on the set-up of the Federal Institute for Materials Research and Testing (Berlin, Germany) with the assistance of researcher Ch. Spitz. The excitation was carried out in the cryostat by the pulse irradiation of the dye laser (R6G dye) with pulse duration 80ps synchronously pumped by pulse mode-locking Ar+ laser. The luminescence of the PIC J-aggregates was separated by monochromator and measured in the photon counting regime by the photo-multiplier.

The measurement of non-linear cubic susceptibilities of J-aggregates' PIC in thin solid films was carried out using the Z-scan method on the set-up shown on fig. 1 based on the dye R6G laser pumped using 5 ns pulsed Nd:YAG laser.

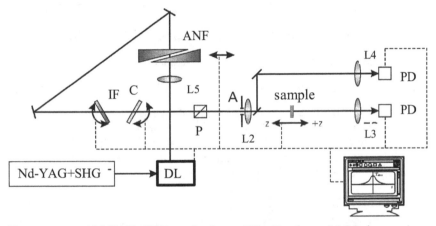

Fig. 1. Z-scan set-up. Nd-YAG+SHG – pulse laser, DL – Dye laser, L1-L4 –lenses, A – Aperture diaphragm, P – Glan prism, IF – interference filter, C– compensator of the beam shift, ANF– adjusted neutral filter, PD–photodiode.

The weak luminescence of the laser dye was cut off by the interference filter IF590 (Carl Zeiss Jena). The light signals were measured by photodiodes and the measured $T(z)$ curves had 200 experimental points with the average value taken from the tens pulses. The calibration was carried out with etalon CS_2 substance ($\gamma=(3.6\pm0.3)10^{-14}$ cm^2/W). The values of the real $Re^{(3)}$ and imaginary $Im^{(3)}$ parts of the cubic susceptibility in the esu units from the measured values' non-linear refraction coefficient γ and absorption coefficient β in SI units were calculated by equations:

$$\mathrm{Re}\,\chi^{(3)} = \frac{cn_0^2}{160\pi^2}\cdot\left(\gamma - \left(\frac{\lambda}{4\pi}\right)^2\frac{\alpha_0\beta}{n_0}\right) \tag{1}$$

$$\mathrm{Im}\,\chi^{(3)} = \frac{\lambda c n_0^2}{640\pi^3}\cdot\left(\beta + \frac{\alpha_0\gamma}{n_0}\right) \tag{2}$$

where, c – light velocity in vacuum, λ - irradiation wavelength.

2.2 The formation of J-aggregates' PIC with the long alkyl chain substituters in thin solid films

The derivatives of PIC with symmetrical and non-symmetrical long alkyl chain substituters were synthesized to stimulate the J-aggregates' formation in thin solid films.

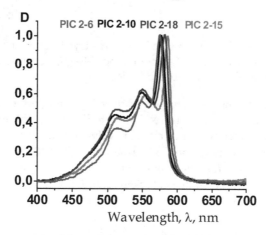

Alk=Alk'=C_2H_5 (PIC2-2),　　　　　　Alk=C_2H_5, Alk'=C_6H_{13}(PIC2-6),

Alk=Alk'=$C_{10}H_{21}$(PIC10-10),　　　　Alk=C_2H_5, Alk'=$C_{10}H_{21}$(PIC2-10),

Alk=Alk'=$C_{15}H_{31}$(PIC15-15),　　　　Alk=C_2H_5, Alk'=$C_{15}H_{31}$(PIC2-15),

Alk=Alk'=$C_{18}H_{37}$(PIC18-18),　　　　Alk=C_2H_5, Alk'=$C_{18}H_{37}$(PIC2-18).

There is no difference in the absorption spectra of the obtained dye in monomer form in organic solution. Symmetrical dyes do not give J-aggregated form in solid films and have the tendency to form H-aggregates. All non-symmetrical dyes with long alkyl chains PIC2-6, PIC2-10, PIC2-15, PIC2-18 give the J-aggregated form in thin solid films, see fig. 2, with maximum at λ=575-583 nm and FWHM 360-400 cm^{-1}.

Fig. 2. The absorption spectra of the non-symmetrical pseudoisocyanines in thin solid films.

There are three electron-phonon transitions (526, 492 и 460 nm) shifted relative to each-other to the quant of the double C=C bound oscillation (1400 cm^{-1}) for the monomer form of dye in the solution. The same three transitions shifted to the long wave side are observed in the structural absorption band with the maximum at 556 nm in the spectra of the non-symmetrical pseudoisocyanines in the thin solid films. This band did not change at the thermal decay of the J-aggregate and belongs to the monomer form of the dye. Thus, the spectrum of dye in solid films is the superposition of monomer and J-aggregated forms.

The absorption in the maximum of the J-aggregates' peak exceeds the absorption in the maximum of the monomer form by as much as 1.5-1.8 times. The relation of the optical density in the J-aggregates' peak and in the monomer maximum (J/M) was used to characterize the degree of the conversion dye to aggregate in thin solid films. The length of the alkyl chain has a weak dependence on the degree of the conversion of the monomer to the J-aggregate in the thin solid films. But the alkyl chain length has a strong influence on the rate and degree of the spontaneous recovering of the J-aggregate after thermal destruction. For example, the J-aggregate of PIC2-2 did not restore at all. The recovering begins after 4 day storing in the box at humidity 80% for the J-aggregate of PIC2-6. The J-aggregate of PIC2-10 restored to 80% after a day of storing and PIC2-15 after 12 hours of storing. The J-aggregate of PIC2-18 is completely restored during 4 hours at room temperature. The increase of the alkyl chain length leads to an increasing degree of restoration and shortening of the time of restoring by order of magnitude. To stabilize the J-aggregates on the substrate, a dye with long alkyl substituters is useful. Such substituters lead to an efficient J-aggregation at room temperature, but the aggregates are unstable upon an increase in temperature above 60°C.

2.3 The spectrophotometry of the thermal decay of PIC J-aggregates with long alkyl substituters

The J-aggregates of the PIC with long alkyl substituters are thermally destroyed with transition to the monomer form. The thermal conversion of the J-aggregates of pseudoisocyanine derivatives is studied by measurement of optical absorption spectra at continuous increase of temperature. The spectral change of the J-aggregates of PIC2-18 in the thin film with temperature increased is shown in fig. 3.

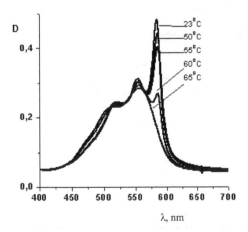

Fig. 3. The spectral change of the J-aggregated film PIC2-18 at sample heating

The J-peak disappears at temperatures above 60ºC and the finished spectrum looks like the broadening spectrum of the monomer form of the dye. The J-aggregates' thermal decay curves are plotted (see fig. 4) and from this it is shown that dyes line to the next row of thermal stability determined on the point of inflection of the thermal decay curve: PIC2-2 (73∘C) > PIC2-6 (67∘C) > PIC2-18 (55∘C) > PIC2-10 (47∘C) > PIC2-15 (37∘C).

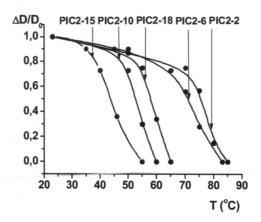

Fig. 4. The thermal decay curves of the J-aggregates of the non-symmetrical dyes in the thin solid films

The effective energies of activation of thermal decay of J-aggregates in thin films were calculated from the non-isothermic curves of the thermal decay. The degree of J-aggregate thermal decay (α_J) is $\dfrac{d\alpha_J}{dt} = k \cdot \left(1 - \alpha_J\right)$

Where, $k = Z \cdot \exp\left(-\dfrac{E_a}{RT}\right)$; Z - preexponential factor, E_a – activation energy, T – temperature.

The differential equation for the degree of thermal decay at constant rate of heating is (Wendlandt, 1974)

$$\frac{d\alpha_J}{dT} = \frac{Z}{\beta} \cdot \left(1 - \alpha_J\right) \cdot \exp\left(-\frac{E_a}{RT}\right) \tag{3}$$

Where, β – the rate of the sample heating (dT/dt). After integration we have

$$\alpha_J = 1 - \exp\left(-\left(\frac{Z}{\beta} \cdot \frac{RT^2}{E_a} \cdot \left(1 - \frac{2RT}{E_a}\right) \cdot \exp\left(-\frac{E_a}{RT}\right)\right)\right) \tag{4}$$

The dependence of the $\left(\ln\left[-\ln\left(1 - \alpha_J\right)\right]\right)$ vs. $1/T$ gives the value E_a from this equation. The experimental and theoretical curves of $\alpha_J(T)$ and double-logarithmic line are shown in fig. 5 for the case of PIC-2-15 J-aggregates' thermal decay.

The obtained values of activation energy for the thermal decay of the J-aggregates of PIC with different long alkyl chains are shown in table 1.

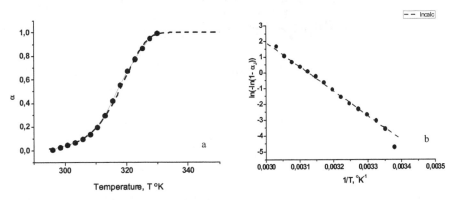

Fig. 5. a,b. Experimental (points) and the model (dash) curves of the degree of the J-aggregates' PIC2-15 thermal decay (a) and its linear approximation vs $1/T$

Dye	E_a kcal/mol
PIC 2-2	43
PIC 2-6	30
PIC 2-10	39
PIC 2-15	30
PIC 2-18	41

Table 1. The values of the activation energy for the thermal decay of the J-aggregates in the thin solid films

We can conclude that the average value of the effective activation energy for the thermal decay of the J-aggregates in the thin solid film is 36 kcal/mol. Let us compare the value of the effective activation energy for the thermal decay of the J-aggregates with the electrostatic energy of the PIC dimer formation. The main contribution to the dimerization energy gives the Coulomb interaction of the two PIC cations and halogen anions (Krasnov, 1984).

$$E_{ion} = -\frac{q_1 \cdot q_2}{r} + A \cdot \exp\left(-\frac{r}{\rho}\right) \tag{5}$$

Where, q_1 and q_2 are charges of ions, r – the distance between ions, A – the coefficient that characterizes the energy of repulsive exchange interactions of the electron orbitals and ρ is the constant equal 0,34 Å. The A value is calculated from the minimum ($\frac{\partial E}{\partial r} = 0$) of the cation-anion interaction.

$$A = -\frac{q_1 \cdot q_2 \cdot \rho}{r^2_{min}} \cdot \exp\left(\frac{r_{min}}{\rho}\right) \tag{6}$$

The value of the energy of the dimer is the minimal at quadruple replacement of the cations and anions as shown in fig. 6. In this case the electrostatic energy of the dimer is

$$E_D = -\frac{4 \cdot q_{PIC} \cdot q_{an}}{r} + 4 \cdot A \cdot \exp\left(-\frac{r}{\rho}\right) + \frac{q^2_{PIC}}{r \cdot \sqrt{2}} + \frac{q^2_{an}}{r \cdot \sqrt{2}} + 2 \cdot A \cdot \exp\left(-\frac{r \cdot \sqrt{2}}{\rho}\right) \qquad (7)$$

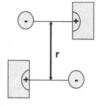

Fig. 6. Quadruple model of the PIC dimer

The effective positive charge of the dye (q_{PIC}~+0,7) and the value of the equilibrium distance cation-anion in the PIC molecule (3 Å) were determined from the quantum chemical calculation using the AM1 method. The map of the calculated electrostatic potential of the PIC iodide is shown in fig. 7.

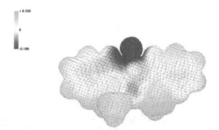

Fig. 7. Electrostatic potential map of the PIC iodide molecule

The value of the dimerization energy was calculated as the difference between the dimer energy and the two molecules PIC in the monomer form.

$$E_D = -\frac{2 \cdot q_{PIC} \cdot q_{an}}{r} + 2 \cdot A \cdot \exp\left(-\frac{r}{\rho}\right) + \frac{q^2_{PIC}}{r \cdot \sqrt{2}} + \frac{q^2_{an}}{r \cdot \sqrt{2}} + 2 \cdot A \cdot \exp\left(-\frac{r \cdot \sqrt{2}}{\rho}\right) \qquad (8)$$

The molecules' distance depending on the dimerization energy is shown in fig. 8.

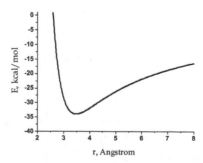

Fig. 8. Distance dependence of the PIC halogen dimerization energy

The minimum of the curve is -34 kcal/mol with the equilibrium distance between PIC molecules 3,5 Å. This is in good agreement with Van der Waals' distance in the dyes' dimers. One can see that the energy of the calculated value of energy of the pair electrostatic interaction of PIC (-34 kcal/mol) is close to the average value of the effective activation energy for the thermal decay of the J-aggregates' PIC (-36 kcal/mol).

The value of the enthalpy of the dyes dimerization in the solution is 6-12 kcal/mol (Ghasemi & Mandoumi, 2008; Coates, 1969; Nygren et al., 1996). It is less than the calculated energy of the dimerization PIC due to the compensatory contribution of the high energy of the dye solvatation. That means that in the solution or in the solvatating polymer the thermal decay of the J-aggregates will happens much more easily than in a non-solvating polymer or in solid films.

2.4 The spectrophotometry of the J-aggregate PIC2-18 iodide to monomer conversion in thin film with the addition of 1-octadecyl-2-methylquinolinium

The number of monomer units in the J-aggregate PIC is an important characteristic in understanding its optical and non-linear optical properties in thin solid films. There are different opinions in the literature about the number of molecular units making up the aggregate: from hundreds (Sundstrom et al., 1988; Tani et al., 1996), tens (Daltrozzo et al., 1974), four (Herz, 1974) or three (Struganova, 2000) molecules in the J-aggregate PIC. Here we describe the difference between the number of molecules which can be in the J-aggregate as in a physical object, for an example in the micelle, and the minimal number of molecules which is enough to give the J-peak in the optical spectrum.

We use another organic cation 1-octadecyl-2-methylquinolinium (MQ18), which has the same charge (+1) and long alkyl tail as PIC2-18, to divide the PIC2-18 molecules in the J-aggregate in a thin solid film. In this way it is possible to determine the number of PIC2-18 molecules in the J-aggregate.

MQ18

The MQ18 is similar to one half of PIC2-18 and statistically replaces the PIC2-18 molecules at the moment of fast J-aggregate assembling during the short time of film formation during spin-coating. In this case the numbers of PIC2-18 molecules in the J-aggregate will depend on the relationship between the concentration of the MQ18 and dye in the film. The probability of PIC2-18 molecule substitution by MQ18 molecule (F_{MQ18}) in the film is equal to the part of the MQ18 molecules in the joint composition PIC18 and MQ18

$$F_{MQ18} = [C_{MQ18}] / ([C_{MQ18}] + [C_{PIC18}]) \qquad (9)$$

The absorption spectra of the films prepared by spin-coating on the basis of mixture PIC2-18:MQ18 (1:0.05,1:0.1, 1:0.2, 1:0.3, 1:0.5, 1:1) were measured and part of them is shown in fig. 9.

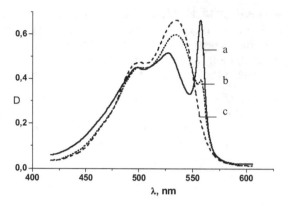

Fig. 9. The film spectra PIC2-18 at MQ18 addition in the mole relation 1:0 (*a*), 1:0.5 (*b*), 1:1(*c*)

One can see the transformation of the J-aggregate spectrum to the monomer dye spectrum with increasing the MQ18 concentration. At the molar relation PIC2-18:MQ18 1:1 we observe only the monomer form of dye. The part of the destroyed J-aggregates (F_J), which is determined as the change of the J-peak optical density (($D^J_0 - D^J_{MQ18}$)/D^J_0) is equal to the probability of the PIC2-18 molecule substitution multiplied by the number molecules (N) needed for J-peak appearance.

$$F_J = (D^J_0 - D^J_{MQ18})/ D^J_0 = N* F_{MQ18} \qquad (10)$$

The value N was calculated from the tangent of the (F_J-F_{MQ18}) dependence shown in fig. 10 is 2.6±0.2. This is between two and three. In accordance with the exciton description of supramolecular assemble (Malyshev, 1993) at the number of molecules in the J-aggregate, at more than two there should be noticeable a hypsochromic shift of J-peak as the aggregate decreases. The J-peak energy depending on the number of molecules in the J-aggregate is:

$$E_k = h\nu + 2V \cos\left(\pi k/(N+1)\right) \qquad (11)$$

where $h\nu$ is the energy of the optical transition of the monomer dye, V – the dipole-dipole interaction of the neighbour molecules in the aggregate, N – the number of molecules in the J-aggregate. The spectral shift for the Lorenz contour of the J-peak (Lr_{ex}) was calculated for the number of PIC molecules in the J-aggregate from 2 to 5 (fig. 11).

$$Lr_{ex} = \frac{\Delta\nu \cdot \dfrac{1}{\sqrt{N}}}{4 \cdot \left(\dfrac{10^7}{\lambda} - \left(\nu_0 - \cos\left(\dfrac{\pi}{N+1}\right) \cdot 2V\right)\right)^2 + \left(\Delta\nu \cdot \dfrac{1}{\sqrt{N}}\right)^2} \cdot N \qquad (12)$$

Where, $\Delta\nu$ -the width of J-peak (300 cm-1), ν_0 -maximum of the J-peak absorption (18770 cm-1), V – 660 cm-1.

The decreasing number of molecules from three to two in the J-aggregate leads to the blue shift of the spectra to 9 nm. In our case the J-peak disappeared, but the blue spectral shift

was not observed. It could be explained in the case of the J-aggregate dimer form presence.
The dimer form of the dye was determined as the main form for the PIC J-aggregate.

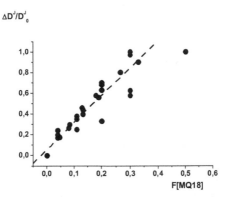

Fig. 10. The relative J-peak optical density change depending on the relative MQ18 content

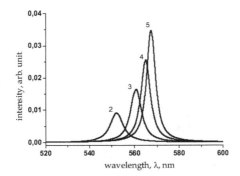

Fig. 11. The absorption spectra of the J-aggregate calculated using eq. 12 for the different
number of the molecules in the J-aggregate (the curve number)

2.5 The optical constants of thin solid films of PIC derivatives

The frequency dispersion of the absorption coefficient k and the refractive index n of the
films of PIC long chain derivatives in J-aggregated and monomer forms were determined in
the wavelength region from 500 to 650 nm. The typical value of measured thin films
thickness was 20-30nm. As an illustration, the resulting spectra for two films are shown in
Fig. 12. The spectra for other films were similar. One shows anomalous dispersion for n
within the region of dye absorption (500-570 nm), with the values $n = 3.05$ and $k = 0.8$ at the
maximum of the J-peak dispersion curve; at the absorption maximum of the J-peak, $n = 2.5$
and $k = 1.25$ for film 21 nm thick.

The value of the index of absorption ($\alpha = 4\pi k / \lambda$) in thin solid films for PIC2-18 is: in the
maximum of anomalous dispersion refraction $\alpha = 2,5 \cdot 10^5$ cm^{-1} for J-aggregate, $\alpha = 1,15 \cdot 10^5$
cm^{-1} in the maximum of absorption for monomer dye ($n_{max}=2.1$). The electron polarizability
for J-aggregated (ρ_J) and monomer forms (ρ_m) of dye were calculated by extrapolation of the

refractive index of dispersion curve to infinite wavelength n_∞ using the least squares method and the Zelmeer equation (Verezchagin, 1980).

$$n_\infty = \sqrt{n - \frac{C}{\left(\lambda^2 - \lambda_0^2\right)}} \qquad (13)$$

Where, C is constant, λ_0 – wavelength of J-peak maximum.

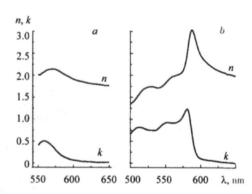

Fig. 12. Refractive indices (n) and absorption (k) of film of PIC 2-18 in the monomeric (a) and J-aggregated (b) form as functions of dispersion measured by spectral ellipsometry

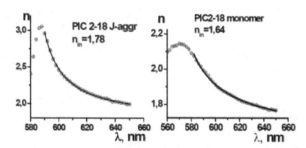

Fig. 13. Dispersion curves for PIC J-aggregate (a) and monomer (b) in the solid film (points) and their least square Zelmeer approximation (line)

The results of the dispersion curves' approximation of J-aggregated and monomer forms of PIC are shown on in fig 13. One can see a good consequence of experimental and fitted curves outside of resonance. The value refractive index for the PIC monomer in film is $n_{\infty m}=1{,}64$ and for J-aggregate is $n_{\infty J}=1{,}78$. The values of the electron polarizability (ρ) is obtained by equation

$$\rho = \frac{3}{4\pi N_A} \frac{n_\infty - 1}{n_\infty^2 + 2} \cdot \frac{M}{d} \qquad (14)$$

Where M is molecular weight (consider dimer value for J-aggregate), d is density of molecules;

gives the values ρ_J=88 Å³ for the J-aggregate and ρ_m=66 Å³ for the monomer form of dye. The electron polarizability of the J-aggregate is 1.33 times more than the polarizability of the monomer. This is not too much to consider the electron delocalization in J-aggregate over the number of dye molecules, but rather it confirms some expansion of electron delocalization due to dye dimerization.

2.6 The luminescent properties of J-aggregate PIC with long alkyl chain

Resonance luminescence takes place for the J-aggregates' PIC in solutions. For example, the PIC forms the J-aggregates in water solution in the presence of phospholipide vesicles with the J-peak of absorption at 580 nm and the J-peak of the luminescence at 582 nm (Sato et al., 1989). The resonance luminescence corresponds to the absence of the molecular coordinate shift between the ground and the excited states' electron energy curves for the S_{00} and S^*_{00} oscillatory states. For the number of samples in solutions, the long wavelength luminescent wing with a maximum of 600-630 nm was observed for J-aggregates at a low temperature (Vacha et al., 1998; Katrich et al., 2000). The authors connect the reason for the long wavelength luminescence with the availability of the dimeric nature traps which accept the excitation energy of the J-aggregate.

The measured steady-state spectra of the luminescence of solid films of J-aggregated dyes PIC2-6, PIC2-10, PIC2-15 at room temperature are shown in fig. 14. The determined Stokes shift for the $v_{00}=1/2(v_{max}^{abs} - v_{max}^{fl})$ transition frequency is 90-100 cm⁻¹ (see table 2). The measured spectra of the monomer PIC luminescence in water:ethanol (1:1) at concentration 10⁻⁴ M/l is shown in fig. 15. The spectrum has the band with broadening counter reaching 700 nm with the maximum at 565 nm. The Stokes shift for the monomer luminescence is 780 cm⁻¹. In accordance with the Lippert approach, the luminescence Stokes shift of the molecule in the medium appears due to the reaction of the medium which is described by the medium function ($f(\varepsilon,n)$) on the dipole moment change at the molecule excitation. The luminescence Stokes shift depends on the polarization function of medium and the square of change of the dipole moment in ground (M_g) and excited state (M_e) divided by the radius of the solvated molecule (R_a) in cube (Levshin & Salecky, 1989; Lakowics, 1983).

$$\Delta E_{st} = 2\frac{(M_e - M_g)^2}{R_a^3} \cdot \left[\frac{(\varepsilon-1)}{(\varepsilon+2)} - \frac{(n^2-1)}{(n^2+2)}\right] \cdot \frac{(2n^2+1)^2}{(n^2+2)^2} \tag{15}$$

The value of the radius of molecule solvate interaction was calculated from

$$R_a = 3\sqrt{\frac{M}{d} \cdot \frac{3}{4\pi N_A}} \tag{16}$$

Where, M, d is the corresponding molecular weight and density of the dye. The obtained value R_a for PIC is R_a=4.5 Å. The calculation of R_a from the molecular volume of PIC cation (14.52x6.41x4.3=400 Å³) gives the same value 4.57 Å. The value the radius of molecular cavity for PIC dimer is R_a = 5.66 Å.

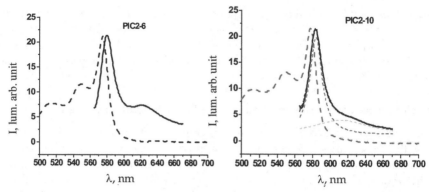

Fig. 14. a,b. The excitation and luminescence spectra in the thin solid films PIC2-6 (a), PIC2-10 (b) with two counter resolution

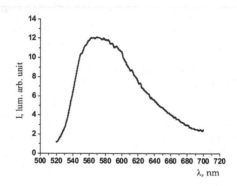

Fig. 15. The luminescence spectrum of the monomer PIC in the water-ethanol solution

Dye,ex. 550 nm	$\lambda_{Макс}$ I lum.,nm	$V_{max.abs}$- $V_{max.lum}$	Stokes shift, ΔV_{0-0}, cm^{-1}	Lorenz counter lum. broadening , cm^{-1}	$\lambda_{Макс}$. 2 lum., nm
PIC2-6	580	150	90	546	624
PIC2-10	583	140	93	502	611
PIC2-15	592	130	102	524	615

Table 2. The luminescent characteristics of the J-aggregates with long alkyl substituters in the thin solid films

The calculation of the dipole moment change for the PIC molecule from the obtained value of the Stokes shift in water-ethanol solution gives the value $\Delta\mu$=2.56D. The analogous calculation for the PIC dimer was carried out in water at Stokes shift 30 cm^{-1} is taken from (Sato et al., 1989) and gives the value $\Delta\mu$=0.7D. One can see from the values of $\Delta\mu$ that the difference of the dipole moments at ground and excited state decreases in 3.66 times with the conversion monomer to J-aggregate form of dye. The calculation of the $\Delta\mu$ change at aggregation PIC at low temperature from the values of the Stokes shift (Renge & Wild 1997) gives the decreasing $\Delta\mu$ at aggregation by 5.5 times. In accordance with the symmetry of the PIC molecule, the dipole moments of the ground and excited state are directed along the short axe of the molecule. It is reasonable to propose that the decreasing of the dipoles moments at the aggregation (dimerization) takes place in the ground and excited state of the aggregate. The decreasing of the dipole moment by 3.7 times is possible at the contra-directed arrangement of the PIC molecules dipoles with the angle between dipoles 43°.

The decreasing of the $\Delta\mu$ apart from the decreasing of the Stokes shift of the luminescence leads to the decreasing of the inhomogeneous broadening of the J-aggregate peak in comparison with the broadening of the dye monomer form absorption band. The square of the inhomogeneous broadening of the molecular spectrum in the medium σ^2_{in} connected with the dipole moment change is proportional to the $\Delta\mu^2/R_a^3$ (Bakhshiev et al., 1989). The decreasing of the $\Delta\mu$ and increasing of the R_a at the dimerization of the dye leads to the decreasing of the inhomogeneous broadening in J-aggregate and the narrowing of the J-peak by 5-7 times in comparison with the spectral width of the monomer form of the molecule. The spectral shift between the maximum of the absorption and luminescence spectra ($\Delta\nu_{a,f}$)

depends on the mean-square dispersion of the inhomogeneous broadening of the luminescence spectra of the molecules in the medium σ^2_{in} that corresponds to half width of the inhomogeneous broadening of the luminescence spectrum.

$$\sigma^2_{\phi\lambda} = \frac{kT}{hc} \cdot \Delta\nu_{a,f} \tag{17}$$

The experimental value ($\nu^{abs}_{max} - \nu^{fl}_{max}$) for J-aggregate PIC2-10 is 140 cm⁻¹ (see table 2). The

calculation of σ^2_{in} for luminescence spectrum of J-aggregates with long alkyl substituters in thin solid films gives a value of full width of inhomogeneous broadening luminescence contour $2\sigma_{in}$=338 cm⁻¹. The comparison of the obtained value $2\sigma_{in}$ (338 cm⁻¹) and experimental value FWHM for luminescence Gauss contour of J-aggregates in thin film 2σ=417 cm⁻¹ allows us to conclude that the luminescence spectrum of the J-aggregates has essential contribution of orthe inhomogeneous broadening makes the essential contribution to the luminescence spectrum of the J- aggregates.

The wide long wavelength shoulder of the luminescence is observed on the all measured samples of the J-aggregated dyes in thin films, apart from the main luminescence peak. The maximum of that luminescence is at 610-625 nm depending on the sample and luminescence slump last to 670 nm. The expansion of the PIC2-10 luminescence spectrum on the two Lorentz contours is shown in fig. 14. On the basis of the measured excitation spectra, it was shown that the long wavelength luminescence does not have its own excitation band and its excitation spectra coincide with the absorption and excitation spectra of the J-aggregate. The long wavelength unstructured wide band in J-aggregated films luminescence we attribute to the exciplex luminescence between J-aggregate and monomer dye or between J-aggregates.

The life time of the exciplex luminescence as a rule is longer than the life time of the luminophore luminescence and has the period of luminescence signal growth relates to the stage of the closing in the molecular contact to form the exciplex. The measurement of the kinetics of the luminescence in the maximum of the long wavelength luminescence 625 nm at room temperature shows the signal growing (see fig. 16) with characteristic time 2.1 ns and luminescence decay with life time 3.6 ns. The luminescence life time in the maximum of the J-aggregate has two components: the short component with life time 20-40 ps and the long component with life time 2.0 ns. One can see that the J-aggregate exciplex luminescence has the period of the signal growing and its life time is more than life time measured in the main peak of the J-aggregate luminescence.

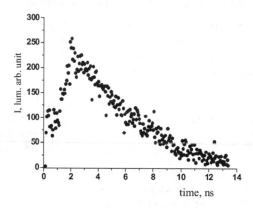

Fig. 16. Kinetics of the luminescence decay of the PIC2-15 at 625 nm in the thin film at the excitation on λ=574 nm

2.7 The formation of J-aggregate PIC in thin solid films. The influence of the cluster anionic derivatives of high boron hydrides

One can see from the map of the electrostatic potential of PIC (fig. 7) that the positive charge of the molecule is distributed in the cavity of electrostatic potential where the anion is situated. From this for the stabilization of the PIC dimer the proper choice is a dianion with dipolar distribution of the negative charge. Cluster dianions of high boron hydrides have the dipolar distribution of the negative charge. The formation of the PIC J-aggregates with the addition of the some cluster high boron hydrides ($K_2B_{10}H_{10}$, $K_2B_{12}H_{12}$, $Cs_2B_{20}H_{18}$, $[Ni^{IV}(1,2\text{-}B_9C_2H_{11})_2]^0$, $Cs[Ni(1,2\text{-}B_9C_2H_{11})_2]$, $NH(CH_3)_3B_9C_2H_{12}$, $NH(C_2H_5)_3[Co(1,2\text{-}B_9C_2H_{11})_2]$, $Cs_2B_{10}H_8I_2$, $[Sn(1,2\text{-}B_9C_2H_{11})]^0$) was studied (Shelkovnikov et al., 2004). It was shown that the addition of anions of high boron hydrides $B_{10}H_{10}{}^{2-}$ and $B_{10}H_8I_2{}^{2-}$ to the PIC2-2 spin-coating solution at molar ratio dye:anion salt 1:0.1 leads to the effective formation of J-aggregates of dye in solid films. The absorption spectra of PIC films doped with $[Ni^{(IV)}(1,2\text{-}B_9C_2H_{11})_2]^0$, $[Sn(1,2\text{-}B_9C_2H_{11})]^0$ and anions $B_{10}H_{10}{}^{2-}$ и $B_{20}H_{18}{}^{2-}$ are shown in fig. 17. The J-peak is distinctly seen in the absorption spectra of the PIC films doped with $B_{10}H_{10}{}^{2-}$ anions. This anion leads to an efficient J-aggregation of PIC with the formation of stable J-aggregates in solid films.

Similarly to the $B_{10}H_{10}{}^{2-}$ anions, the addition of the neutral complex of nickel stimulates the predominant formation of J-aggregates of PIC in a solid film (J/M=1.52). The addition of the $B_{10}H_8I_2{}^{2-}$anion at holding times prior to centrifugation of more than 10 min leads to the formation of stable J-aggregates (J/M = 1.46). At short holding times prior to centrifugation (less than 5 min), a long wavelength shoulder located at 625 nm and extending up to 800 nm was observed in the absorption spectra (Fig. 18). In this case, the optical density of the J-peak decreases. When a film doped with $B_{10}H_8I_2^{2-}$ anions is applied to a glass surface treated with a silicon-containing surfactant for the surface wettability improvement, a high conversion of the monomer to J-aggregates ($J/M = 2.2$) is observed (Fig. 18, spectrum 3).

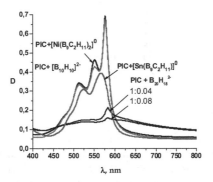

Fig. 17. Absorption spectra of PIC films doped with (1) $[Ni^{IV}(1,2\text{-}B_9C_2H_{11})_2]_0$, (2) $[Sn(1,2\text{-}B_9C_2H_{11})]^0$, (3) $B_{10}H_{10}^{2-}$ and (4, 5) $B_{20}H_{18}^{2-}$. Dye–anion mole ratio: (4) 1 : 0.04; (5) 1 : 0.08

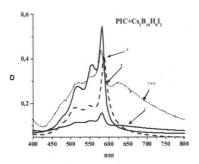

Fig. 18. Absorption spectra of PIC films doped by $B_{10}H_8I_2^{2-}$ anions. The holding time prior to centrifugation: (1) 2–3 min; (2, 3) longer than 10 min. Sample 2 on a pure glass substrate; sample 3, a glass substrate treated with a surfactant

A high stability of the J-aggregates of PIC-closo-hydrodecaborate allowed us to perform an additional doping of the dye film with organic cation salts without destroying the J-peak. The following salts of organic cations were used: methylcholine iodide, tetrabutylammonium (TBA) iodide, laurylcholine iodide and dodecylpyridine iodide. The salts of organic cations were introduced into the films at a mole ratio of the dye cation:added cation varying from 1 : 0.5 to 1 : 5. Upon an increase in the concentration of an organic cation to the mole ratio dye–cation 1 : 4, the J-peak increases in intensity and narrows by 1.5–2 times, as is shown in Fig. 19 for the case of TBA. At a higher concentration of organic cations in the film, the J-peak decays. The destruction of the J-aggregates was observed only upon a high dilution of a dye film with organic cations. This demonstrates the high stability of the J-aggregated form of PIC-closo-hydrodecaborate. For example, the addition of octadecyl choline iodide to films of PIC iodide with long alkyl substituters ($C_{18}H_{37}$) leads to the decomposition of the J-aggregates to the monomer at a dye–additive mole ratio even as low as 1 : 0.1. In fact, at a concentration of the additive organic cation at 4 times higher than the concentration of the dye cation, the molecules of the organic salt of the additive cation can be considered as a kind of a matrix for the dye molecules.

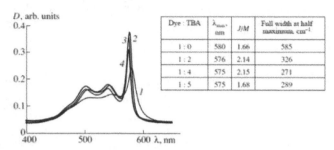

Dye : TBA	λ_{max}, nm	J/M	Full width at half maximum, cm^{-1}
1 : 0	580	1.66	585
1 : 2	576	2.14	326
1 : 4	575	2.15	271
1 : 5	575	1.68	289

Fig. 19. Absorption spectra of films of PIC-closo-hydrodecaborate doped with tetrabutylammonium iodide (the dye–additive mole ratios, maximum wavelengths of the J-peak, J/M ratios and the full widths of the peaks at half maximum are given in the inset). Spectra 1–4 correspond to the dye–additive mole ratio varied from 1 : 0 to 1 : 5, respectively

The dilution of a film of PIC-closo-hydrodecaborate with organic cations leads to an increase in the luminescence of J-aggregates. The luminescence spectra of the J-aggregates in a film of PIC-closo-hydrodecaborate and in an analogous film diluted with TBA cations in the ratio dye–TBA 1 : 4 measured under identical excitation conditions are shown in Fig. 20. It is seen from this figure that, upon dilution of the J-aggregated film by TBA cations, the luminescence intensity of the J-aggregates increases by an order of magnitude. This effect can be explained by a decrease in the concentration quenching of luminescence of the J-aggregates.

In our opinion, the changes in the shape of the J-peak in the absorption and luminescence spectra of thin films of PIC-closo-hydrodecaborate observed upon dilution of the films with foreign organic cation can be explained by the fact that this peak arises due to the absorption of strongly coupled PIC dimers. A stable dye dimer may form because of a strong electrostatic interaction of two cationic molecules PIC+ with a doubly charged polyhedral $B_{10}H_{10}^{2-}$ anion. In a dimer PIC_2-$B_{10}H_{10}$, the electrostatic interaction is saturated and, upon dilution with foreign cations, large-size aggregates should decompose.

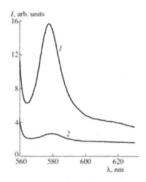

Fig. 20. Thin solid films luminescence spectra (1) PIC-closo-hydrodecaborate doped with TBA at a dye–TBA mole ratio of 1 : 4 and (2) pure PIC-closo-hydrodecaborate

The charge distributions in the dye cation and in the anions of the highest boron hydrides
were calculated by the AM1 method and are shown on the structures below. Qualitatively,
the calculated charge distribution among the atoms of the known anions $B_{10}H_{10}^{2-}$ and
$B_{12}H_{12}^{2-}$ agrees with the charge distribution calculated by such methods as PRDDO, 3D
Huckel theory and STO-3G (Pilling & Hawthorne, 1964). Quantitatively, the best agreement
was obtained for the total charges of the B–H groups.

In the $B_{10}H_{10}^{2-}$ anion, a negative charge is distributed non-uniformly. The vertex B–H groups
carry a larger negative charge (–0.347) than the equatorial B–H groups (–0.163). In the PIC
cation, a highest positive charge is located on two carbon atoms of the heteroaromatic rings
involved in the chain of conjugation. The remaining carbon atoms and nitrogen heteroatoms
carry negative charges. The remaining positive charge is mainly distributed over the
hydrogen atoms of the aromatic rings. The basic atomic centres of the electrostatic attraction
of the ions considered are the vertex B–H groups in the $B_{10}H_{10}^{2-}$ anion, the carbon atoms and
the hydrogen atoms of the heteroaromatic rings of the dye cation. The geometrical
dimensions of the $B_{10}H_{10}^{2-}$ anion (the distance between the polar hydrogen atoms is equal to
(5.9 Å)) correspond well to the dimension of the cavity of the electrostatic potential of PIC
due to the turn of its aromatic planes, in particular, the distance between the positively
charged hydrogen atoms H1 and H2 (5.6 Å). It should be noted that the involvement of the
vertex B–H groups of the $B_{10}H_{10}^{2-}$ anion into the formation of a stable PIC-closo-
hydrodecaborate complex is confirmed by the data of IR spectroscopy and x-ray diffraction
analysis (Cerasimova, 2000). The structure of a dimer of PIC-closo-hydrodecaborate
calculated using the MM+ method and the map of the electrostatic potential calculated
using the semi-empirical quantum-chemical AM1 method are shown in Fig. 21. It is seen
that the aromatic planes of the dye cations interacting via the anion of closo-
hydrodecaborate are turned with respect to each other. This turn can be explained by the
mutual repulsion of the positively charged aromatic hydrogen atoms in the PIC dimer.

The charge distribution in the $B_{10}H_8I_2^{2-}$ anion is more polarized than in the $B_{10}H_{10}^{2-}$ anion.
The negative charge is strongly displaced toward the vertex boron and iodine atoms. If one
assumes that the dipole charge distribution in the $B_{10}H_{10}^{2-}$ anion facilitates the formation of
J-aggregates of PIC, then, upon addition of $B_{10}H_8I_2^{2-}$ anions, efficient J-aggregation should

also be observed. This agrees with the formation of J-aggregates of PIC in the presence of these anions in the case of the solution being held prior to its centrifugation for a time period longer than 10 min (Fig. 18). Upon deposition of a dye solution immediately after the addition of anion, an intensive association of the dye is likely to occur, which is stimulated by a relatively high initial concentration of anions localized near the drop of salt addition (later, the associates gradually dissolve). In this case, the intensive association of the dye leads to the formation of a set of strongly aggregated dimers of PIC, whose absorption extends toward the long wavelength range up to 800 nm. The maximum of this distribution (625 nm) possibly corresponds to the dye tetramers.

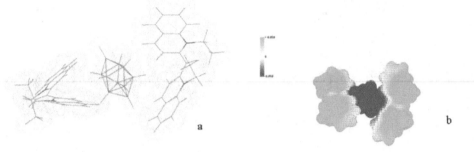

a   b

Fig. 21. The calculated structure of the PIC-closo-hydrodecaborate dimer (a) and dimer electrostatic potential map (b).

In the $B_{12}H_{12}^{2-}$ anion, the charge distribution among the boron atoms is completely uniform and the absolute value of the charge of the B–H groups (–0.167) is significantly smaller than that of the vertex B–H groups of the $B_{10}H_{10}^{2-}$ anion. Note that this anion has no poles of concentration of a negative charge and, therefore, no structure-forming effect on the formation of J-aggregates of PIC. Taking into account the charge distribution pattern in the considered cluster anions of boron hydrides, as well as their relative dimensions, we can conclude that a more pronounced bipolar distribution of the negative charge and a smaller volume of an anion facilitate the formation of a J-aggregate of PIC.

In the case of a strongly bound dimeric form of PIC-closo-hydrodecaborate, the dissociation of these dimers to the monomer upon dilution of dye films with a salt of an organic cation does not occur immediately, but rather via the stage of formation of separated dimers of PIC. The weakening of the intermolecular interaction leads to (i) a narrowing of the absorption peak of a J-aggregate, (ii) an increase in the intensity of the J-peak, (iii) a hypsochromic shift of the J-peak and (iv) an increase in the intensity of luminescence of the J-aggregate. We assume that the dimer of a dye is the main structural unit responsible for the absorption of a J-aggregate (J-peak) of PIC.

2.8 The thermal stability of PIC-closo-hexahydrodecaborate in the thin solid films

The thermal stability of the J-aggregates of PIC with long alkyl substituters in the thin solid films (T_{decay} about 60° C) is near the temperature of the film heating at pulse laser excitation. The laser destruction of these films takes place at laser intensity I_0=1-5 MW/cm². The film temperature change (ΔT) at the named intensity is 20-90° C from the equation:

$$\Delta T = \frac{I_0(1-10^{-D}) \cdot (1-R) \cdot \tau}{h \cdot \rho \cdot C_p}$$ (18)

where, ρ - density 1,4 g/cm³, C_p – specific heat capacity 3,1 J/g·K⁰, D – film optical density
0,5, R – reflection coefficient 0,3, laser pulse duration τ=5ns, film thickness h=30nm.

The J-aggregates of PIC-closo-hexahydrodecaborate in the thin solid films are more suitable
for the non-linear laser experiments due to their higher thermal stability compared with that
for the J-aggregates of long chain PIC.

The thermal stability of PIC-closo-hexahydrodecaborate with TBA addition in solid films is
studied by using the spectrophotometry method. The J-aggregates of PIC-closo-
hexahydrodecaborate are stable at sample heating to temperature 90⁰C and being partially
destructed at higher temperature are restored to 100% at sample cooling to room
temperature. As an example the thermal decay and restoring of the J-aggregates' PIC-$B_{10}H_{10}$
the spectral change of the J-aggregated film with addition of the TBA for peak narrowing is
shown in the fig. 22 a. One of the features of the thermal decay of the J-aggregates' PIC in
the thin solid films is the red shift of J-aggregate peak. An example of an evident red shift of
absorption peak of J-aggregates is shown in fig. 22 b,c.

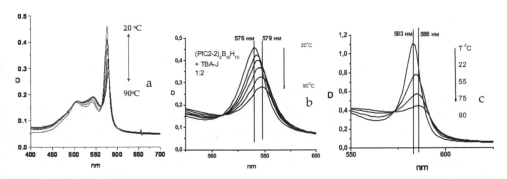

Fig. 22. The absorption spectra change of the J-aggregated films' PIC-closo-hydrodecaborate
headed (cooled), doped with TBA at a dye–TBA mole ratio of 1 : 4 (a), expanded scale (b),
expanded scale PIC2-2 in polymer anethole (c)

The red shift of the J-aggregate peak in thin solid films and polymer matrix to about 3 nm
was observed in the causes of J-aggregates' thermal decay. This effect we connect with the
inhomogeneous broadening of the J-aggregate absorption band. The red spectral shift
caused by intermolecular interaction follows the depth of the J-aggregate potential in the
ground state for the homogeneous absorption profile as it is considered for the positive
solvatochromic shifts of the absorption spectra of the molecules (Suppan, 1990). The thermal
decomposition takes place for local aggregates having different activation energy. The
activation energy has inhomogeneous Gaussian distribution between J-aggregates sites with
homogeneous spectral Lorentz profiles. In this case, each homogeneous Lorentz profile has
the spectral position with proper own activation energy of thermal decay from aggregate

(dimer) to monomer. At the sample heating the aggregates with lower activation energy (ΔE_l) have a higher probability to destroy and have higher excitation energy. The inhomogeneous distribution of J-aggregates leads to spectral-kinetic non-equivalence of the thermal decay. The statistic weight of J-aggregates with high energy of activation increases during thermal decay.The combination of inhomogeneous broadening of initial homogeneous spectral Lorentz profiles of absorbing aggregates gives the Voight function (V) of spectral distribution.

$$V = 2\ln(2)\frac{\gamma}{\Delta w^2} \cdot \int_{-\infty}^{\infty} \frac{e^{-t^2}\, dt}{\left(2\ln(2)\cdot\frac{w-w_0}{\Delta w} - t\right)^2 + \left(\ln(2)\frac{\gamma}{\Delta w}\right)^2} \tag{19}$$

Where, γ – width of Lorentz profile, w - width of Gauss profile and w_o – central position of absorption band. To take into account the individual rate of thermal decomposition of absorbing Lorentz profiles of aggregates, we need to include in the Voight function the equation describing the thermal decay curve, depending on the energy activation distribution. The equation for thermal decomposition as follows from (4) shown in the above curve is

$$1 - \alpha_T = \exp(-F_a)$$

$$F_a = \left[\frac{T^2 R}{\Delta E_a}\cdot\left(1 - \frac{2TR}{\Delta E_a}\right)\cdot\exp\left(-\frac{\Delta E_a}{TR}\right)\right]\cdot Z \tag{20}$$

Where, α_T is the degree of J-peak decay, T – temperature $^{\circ}$K, R – universal gas constant, Z – preexponential factor Z=10^{18} and ΔE_a – activation energy with average value 120 kJ/mol. The values correspond to the thermal decomposition behaviour of J-aggregates' peak for PIC2-15 in thin films. We suppose that the activation energy spectral distribution has the normal distribution profile in spectral region of J-aggregate absorption with deviation ±10 kJ/mol. The Voight function, including the degree of thermal decomposition $V(E_a, T)$, is

$$V(E_a, T) = 2\ln 2 \cdot \frac{\gamma}{\Delta w^2} \int_{-\infty}^{\infty} \frac{e^{-t^2}\cdot e^{-F_a}}{(2\ln 2\frac{w-w0}{\Delta w} - t)^2 + (\ln(2)\cdot\frac{\gamma}{\Delta w})^2}\, dt \tag{21}$$

Calculation of the Voight contour of J-peak absorption was carried out. The values of homogeneous and inhomogeneous broadening of the J-peak are approximately equal and make up 235 cm^{-1} at room temperature. The results of modelling of the Voight contour of J-aggregate PIC2-15 absorption at the stage of thermal decomposition was compared with the experimental spectral curves of J-aggregate (see fig. 23).

One can see good agreement between modelling and experimental spectral curves. The results confirm the influence the inhomogeneous broadening of absorption spectra of J-aggregates in thin films on the red shift of absorption band of J-aggregates at thermal decay.

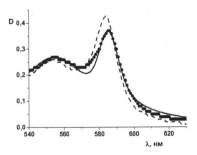

Fig. 23. Calculated Voight contour of J-aggregates at 55°C (solid lines) and experimental absorption curves of thin solid films of J-aggregates of PIC2-15: dash curve - initial spectrum, dot curve - spectrum at 55°C

2.9 The spectrophotometry of J-aggregate thin film formation by spin-coating

Thin film formation of J-aggregated pseudoisocyanine iodide with long alkyl substituents PIC 2-18 and PIC 2-2 with the addition of $B_{10}H_{10}^{2-}$ anion was studied by spectrophotometry directly during spin-coating of the dye solutions. The view of spectra change during spin-coating of a solution of PIC 2-18 at 2000 rpm from the time of placing the solution until formation of the solid J-aggregated film is shown in Fig. 24. The inset shows the change of optical density at the absorption maximum of the monomer form of dye Fig. 24a. The initial spectrum of dye solution on the substrate exhibits optical density noise fluctuations because it exceeds the dynamic range of the spectrophotometer (2.5D). After 2.5 sec from the spin-coating stars, the dye absorption in solution stabilizes at optical density $D=1.3$ and remains unchanged for 0.4 sec. This means that the centrifugal discharge of dye solution from the substrate is finished and the film is held on the substrate by surface-tension forces. The dye concentration increases because of solvent evaporation from the thin film, whereas the number of dye molecules on the substrate remains the same. Fig. 24b shows the spectral changes for a film of PIC 2-2 (+$K_2B_{10}H_{10}$) at rotation rate 1500 rpm. The time for formation of J-aggregates on substrate is 0.45 sec.

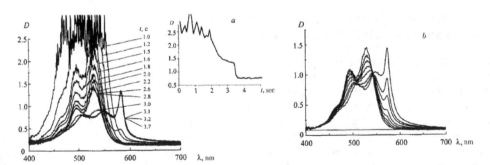

Fig. 24. Spectral changes during spin-coating of solutions of pseudoisocyanine derivatives: PIC 2-18 solution [the time from the moment of placing the solution on the centrifuge is shown; in the inset, optical density at the absorption maximum of the monomeric dye in solution (530 nm) as a function of spin-coating time] (a); PIC 2-2 solution with added $K_2B_{10}H_{10}$ during formation of J-aggregate (spectra recorded every 0.05 sec) (b).

It should be noted that the experimental spectra are not corrected for double passage of light through the sample. Losses of light to double reflection and absorption of dye film must be considered in order to obtain the true value of the optical density in our optical setup. For this, the reflectance spectrum of J-aggregated PIC 2-18 film was measured (Fig. 25). The corrected optical density of J-aggregated film $D(\lambda_{cor})=\log\left[I_0/(I_0-A-R)\right]$ was calculated by solving the quadratic equation for the absorption coefficient obtained from the formula for the measured optical density $D(\lambda)$:

$$10^{D(\lambda)} = \frac{I_0}{I_0 - A - R - R(I_0 - A - R)(I_0 - A - R)},$$

$$A^2 - A(1 - 2R + I_0) - \left(\frac{I_0}{10^{D(\lambda)}} - I_0 + R + RI_0 - R\right) = 0. \tag{22}$$

where A is the absorption coefficient and R the film reflectance coefficient for each wavelength.

A comparison of Fig. 24 and Fig. 25 shows that correction of the spectrum taking into account double absorption and reflection decreases significantly the optical density of the sample at the maximum of the J-peak (to 0.3).

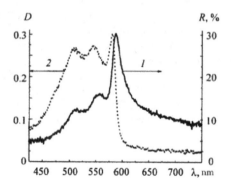

Fig. 25. Reflectance spectrum (1) and absorption spectrum calculated taking into account reflectance of the sample (2), J-aggregated film of PIC 2-18

Spectral changes occurring during spin-coating of J-aggregated PIC dye are related to the increased refractive index of the medium at the time of film formation. The bathochromic shift of the monomer dye absorption maximum is due to solvent evaporation and replacement of the initial medium with refractive index n_1, consisting of solvent molecules, by a medium with refractive index $n2$, consisting of dye molecules themselves. The absorption maximum of the dye for PIC film (+$B_{10}H_{10}{}^{2-}$) shifts from 529 to 550 nm as the film dries out; for PIC 2-18 film, from 529 to 551 nm. The spectral shift in organic molecules on going from vacuum to a medium with refractive index n_0 is described by the universal interaction function (Bakhshiev et al., 1989)

$$\Delta v = \text{const} \cdot \frac{n_0^2 - 1}{n_0^2 + 2} \tag{23}$$

An empirical expression for the universal interaction function of dye with medium was obtained based on experimental data for absorption of PIC monomer in various media (Renge & Wild 1997)

$$\lambda(n_0) = \frac{10^7}{19716 - 2964\left[\left(n_0^2 - 1\right)/\left(n_0^2 + 2\right)\right]} \tag{24}$$

where $\lambda(n_0)$ is the wavelength (nm) of the absorption maximum for dye in a medium with refractive index n_0.

The refractive index of the medium for dye solution is $n_m=1.405$. The wavelength of the absorption maximum of monomer dye λmax=526.6 nm was calculated using Eq. (24) and agrees with the experimental value for a dilute dye solution. The refractive index of film after centrifugal discharge of dye solution from the substrate obtained using Eq. (24) was 1.47 for absorption of PIC monomer at 529 nm. From this point the spectral shift of the absorption band of dye monomer during spin-coating from 529 nm to 551 nm corresponds to a transition from a medium with refractive index n=1.47 to a medium with $n=2.1$. That is in agreement with n value from the dispersion curve for dye in thin film (see fig. 12a).

Thus, the bathochromic shift of the dye absorption spectrum at the time of film formation is due to the increased polarizability of the medium. Therefore, the polarization spectral shift must be considered in determining the exciton splitting of the monomer absorption band upon J-aggregation or the energy of exciton coupling of dye molecules in the J-aggregate (Jex). The experimental value Jex is determined from the energy difference of the long wavelength transition of PIC in solution (530 nm) and in the J-aggregate using the formula (Kuhn & Kuhn 1996)

$$\Delta E = \sqrt{E_m^2 + 4E_m J_{ex}} - E_m \tag{25}$$

where, Em is the energy of the long wavelength transition in the monomer; Jex, the exciton coupling energy; and ΔE, the energy difference of absorption at the monomer peak and the J-peak. The absorption maximum of the J-peak was observed at λ=570 nm where the absorption of PIC J-aggregates was measured in aqueous ethyleneglycol matrix at liquid helium temperature (Minoshima et al., 1994). In this instance Jex=632 cm-1. A similar value of Jex is used in theoretical calculations. In particular, Jex $\approx$ 600 cm-1 is used to determine the exciton delocalization length (Bakalis & Knoester, 2000).

The change of local field for a dye molecule in the aggregate as a result of the change of medium is not considered for this estimate of Jex. The total energy of the J-peak shift of aggregate absorption is the sum of the polarization energy and the exciton coupling energy of neighbouring molecules in the aggregate. Therefore, the bathochromic shift of the dye S_{00}-transition that is due to the medium polarizability with an effective refractive index equal to the refractive index of the dye monomer in the solid phase, i.e., $n=2.1$, must be taken into account for dye molecules in the aggregate. The exciton coupling energy is Jex=293 cm-1 for the spectral maximum of monomer absorption shifted to 555 nm and for the absorption maximum of the J-peak located at 574 nm. Thus, the resulting value Jex $\approx$ 300 cm-1 is half of that used in the literature (Jex $\approx$ 600 cm-1). The values Jex for aggregates of other dyes, known from the literature (Moll, 1995), are also estimated based on the spectral shift of the absorption maximum of J-aggregated dye relative to the monomer in solution. The polarization spectral shift for dyes with highly polarizable π-electrons is significant. This correction must be considered in estimating Jex, not only for PIC aggregates, but also for aggregates of any other dyes.

Another aspect of the change the local field factor at the J-aggregate formation is the dramatic increase of the J-peak at the last moment of formation of the solid film without a decrease of the optical density at the absorption maximum of the monomeric dye, i.e., the significant growth of the optical density of the J-peak is not compensated by a decrease of the monomer optical density. An additional increase of the J-peak, for example, for PIC 2-2 ($+K_2B_{10}H_{10}$) by 2.15 times can be seen from the measured absorption spectra.

One reason for the increased absorption as the film dries out is the increased refractive index of the medium. According to Bakhshiev et al. (1989), the ratio of integrals of absorption intensity (Int) in media with different polarizabilities (Int_{max} for n_{max} and Int_{min} for n_{min}) depends on the local field factor

$$\frac{Int_{max}}{Int_{min}} = \frac{\left(n_{max}^2 + 2\right)^2}{9n_{max}} \frac{9n_{min}}{\left(n_{min}^2 + 2\right)^2} \tag{26}$$

The monomer concentration ceases to decrease with a spectral shift of the monomer peak to 539 nm. According to Eq. (24), the refractive index of the medium at this moment $n=1.7$. The ratio of local field factors is 1.94 for a change of medium refractive index from 1.7 to 2.5. The calculation in the optical density change, taking into account the reflection and absorption growth due to growth of the local field factor, gives the increasing measured optical density of the J-peak by 2.2 times and explains the increase of the J-peak without loss of dye monomer.

The results from the investigation of J-aggregate formation during spin-coating of PIC indicate that the change of local field factor should be considered in interpreting spectral properties of nano-structured aggregates for any type of dye aggregation or for other highly polarizable molecules in both films and solutions.

2.10 The third order non-linear optical properties of pseudoisocyanine J-aggregates in thin solid films

The third order non-linear optical properties of pseudoisocyanine J-aggregates in thin solid films were studied using the Z-scan method as well for PIC iodide anion as for PIC with closo-hexahydrodecaborate anion ($B_{10}H_{10}^{2-}$) (Markov et al., 1998a, 1998b; Plekhanov et al., 1998a, 1998b). The dispersions values of imaginary $Im\chi^{(3)}$ and real $Re\chi^{(3)}$ parts of cubic susceptibility of J-aggregated film are shown in fig. 26. In both causes the non-linear bleaching of the solid film samples in J-peak maximum is observed at resonance laser irradiation excitation ($I>10^5$ W/cm^2) and obvious darkening on the red side of the dispersion curve of imaginary part of $\chi^{(3)}$. J-aggregates of both types of dye have a similar value of the coefficient of non-linear absorption $\beta \approx -6*10^{-2}$ cm/W and corresponding value $Im\chi^{(3)} \approx -1*10^{-5}$ esu at the bleaching maximum and $Im\chi^{(3)} \approx 0.12*10^{-5}$ esu at the maximum of darkening. At the intensity of the laser irradiation at more than 3 MW/cm², the films become irreversibly burned.

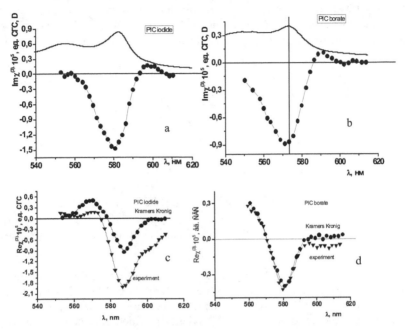

Fig. 26. a-d. The dispersion curves of the imaginary part of the cubic susceptibility of the J-aggregated $Im\chi^{(3)}$ PIC-iodide (a) and closo-hexahydrodecaborate (b) and real $Re\chi^{(3)}$ PIC-iodide (c) and closo-hexahydrodecaborate (d) compared with Kramers-Kronig calculated curves

The maximal value of the real part of the cubic susceptibility for films with J-aggregate of PIC-iodide is $Re\chi^{(3)}=-1,8*10^{-5}$ esu and with J-aggregate of PIC-closo-hexahydrodecaborate $Re\chi^{(3)}=-0,45*10^{-5}$ esu. This difference is due to the contribution of the refraction thermal change in the higher absorption PIC-iodide film compared with absorption of PIC-$(B_{10}H_{10}^{2-})$ film as is shown in fig. 27. The imaginary part of the cubic susceptibility does not depend on the thermal contribution. The sample refraction change caused by saturation of the electron transition without the part of thermal contribution was calculated by Kramers-Kronig relation connecting the $Im\chi^{(3)}$ and $Re\chi^{(3)}$ for the non-linear case (Markov et al., 1998c) (see fig. 26 a-d).

$$Re\,\chi(\Omega) = \frac{1}{\pi}\int_{-\infty}^{\infty}\frac{Im\,\chi(\Omega)}{x-\Omega}dx - 2\,Re[C^{-1}_{(\Omega_p)}(\Omega)] \tag{27}$$

where C^{-1} is the residue in the pole of the $\chi(\Omega)/x-\Omega$, $(\Omega=\omega-\omega_j)$ function.

The thermal contribution to the non-linear refraction for the films with the optical density $D<0.4$ is insignificant, but it should be considered for films with optical density $D\approx1$. The maximal values of the real parts of $\chi^{(3)}$ with and without taking into account the thermal contribution are close to each other and are $Re\chi^{(3)} \sim -(0.5-0.7)*10^{-5}$ esu. The obtained values of the non-linear response of PIC J-aggregates in the thin solid films ($\chi^{(3)} \approx 10^{-5}$ esu) are two orders of magnitude higher than in the water solution (Bogdanov et al., 1991) and polymer matrixes (Plekhanov et al., 1995) ($\chi^{(3)} \approx 10^{-7}$ esu). Apart from the much lower values of the cubic susceptibility, there was no observation of non-linear darkening of PIC J-aggregates.

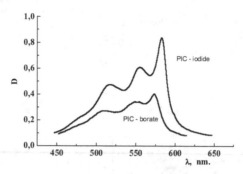

Fig. 27. The absorption spectra of the PIC2-2 iodide and closo-hexahydrodecaborate J-aggregated films

The iodine anion existing in the PIC molecule has the high polarizability and the more the iodine atoms are in the anion molecule, the higher the polarizability. The PIC J-aggregated film with addition of high polarizability tetraethylammonium salt of the closo-tetrahydrohexaiodo-dodecaborate ($B_{10}H_4I_6^{2-}(C_4H_9)_4N^+$) which has six iodine atoms was prepared to raise the J-aggregate surrounding polarizability. The dispersion curve of the imaginary part of the obtained film is shown in fig. 28 compared with the dispersion for the PIC-closo-hexahydrodecaborate film.

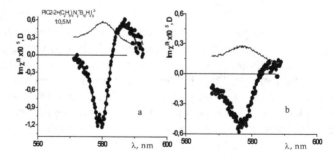

Fig. 28. The cubic susceptibility dispersion curves for the PIC J-aggregated films with - $B_{10}H_4I_6^{2-}$ 1:0,5 (a) and $B_{10}H_{10}^{2-}$ 1:0,5 (b)

As one can see from the adducing figures, the non-linear optical response of the J-aggregated film samples with high polarizable anion ($B_{10}H_4I_6^{2-}$) is higher than for the film with $B_{10}H_{10}^{2-}$ anion. The negative value of the imaginary part of the cubic susceptibility increases 2 times and the red shifted induced non-linear absorption increases in 5 times up to $0.6 \cdot 10^{-5}$ esu. That means it is possible to act upon the non-linear response of the charged aggregated molecules by additions of the proper anions.

The non-linear darkening on the red side of dispersion $Im\chi^{(3)}$ curves is considered on the basis of the four-level model of excitation energy relaxation in PIC J-aggregates shown in fig. 29. The additional level of the relaxation of the first singlet state of J-aggregates in the thin film arises from high polarization of the surrounding J-aggregate medium consisting of the PIC dye molecules. The equations obtained in (Tikhonov & Shpak, 1977) for a four-level model of the excitation relaxation in assumption $k_4 \gg k_2$ were used to calculate the induced bleaching and darkening change in the J-aggregated thin film depending on the intensity of the incident laser radiation.

$$\Delta T = \frac{1 + \dfrac{I_0 \sigma_{32}}{A_k} \cdot \left(\dfrac{1}{\tau_2} - \dfrac{\sigma_{10}}{\tau_3 \sigma_{32}} \right)}{1 + \dfrac{1}{A_k} \cdot \left[\sigma_{10} I_0 \left(\dfrac{1}{\tau_2} + \dfrac{1}{\tau_3} \right) + \sigma_{10} \sigma_{32} I_0^2 \right]} - 1 \tag{28}$$

$$A_k = \left(\frac{1}{\tau_1} + \frac{1}{\tau_2} \right) \cdot \left(\frac{1}{\tau_3} + \sigma_{32} I_0 \right) \qquad \sigma = \frac{2303 \cdot \varepsilon}{6.02 \cdot 10^{23}}$$

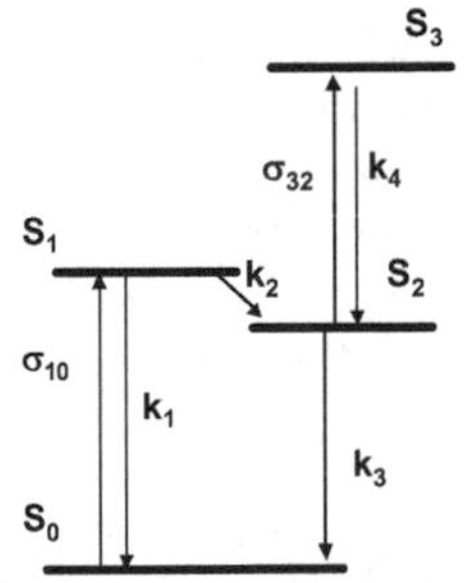

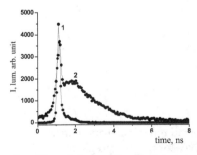

Fig. 29. Four-level model of the J-aggregate excitation relaxation in thin film, where, S_2 excited state is considered here as relaxed S_1 state

The luminescence kinetics decay for J-aggregated film was measured to estimate the values of the characteristic times of excited state relaxation. The luminescence decay of the J-aggregated thin film in the maximum of the J-aggregate luminescence $\lambda = 590$ nm at the excitation in J-peak maximum $\lambda = 574$ nm at temperature 94^0K is shown in fig. 30.

Fig. 30. Signal instrument function (1) and decay kinetics luminescence curve J-aggregates PIC2-15 in the film (2) excitation $\lambda = 574$ nm measurement at 590 nm, T=96^0K

The fast and slow components of the luminescence decay take place on the luminescent decay kinetic curve. The signal instrument function was 80 ps. The expanding of the instrument function due to the appearance of the fast component is 20-40 ps. The life time of the slow component is 2 ns. The period of the luminescence growth of the slow component takes place with characteristic time 0.1-0.2 ns. The existence of the fast and slow components is typical for the dye molecules in polar liquid and solid phases. Some relaxations times from picoseconds to nanoseconds relate to different processes of the solvate shell relaxation depending on the characteristic times of the anionic and cationic parts of the medium relaxation in the case of organic or inorganic salts (Ferrante et al., 1998; Saha et al., 2004; Das et al., 1996; Mandal et al., 2002; Arzhantsev et al., 2003). The luminescence of the J-aggregate in the thin solid film occurs from the two states: the state of the fast matrix relaxation and state of the slow matrix relaxation which is populated in a time of about 100 ps. Different luminescent life times 20-40 ps, 128, 659 пс, 1.7 нс were reported depending on the preparation condition and irradiation intensity (Sundstrom et al., 1988) for the PIC J-aggregates in solvents.

The initial values of the parameters of the model are determined as: J-aggregate extinction coefficient $\varepsilon_{10}=2\ 10^5$ $M^{-1}см^{-1}$, the same value for S_2-S_3 transition $\varepsilon_{32}=2\ 10^5$ $M^{-1}см^{-1}$, $\tau_2=100$ ps (from the time of the slow component luminescence increasing), $\tau_3=2\ 10^{-9}$s (the life time of the luminescence decay slow component), $\tau_1=10^{-11}$s is a typical value of the non-radiation relaxation in organic molecules, the relation of the width values of the transition S_{1-0} and S_{3-2} spectral contours was taken from measurement of dispersion of the non-linear bleaching and darkening as 150 and 120cm^{-1}. The model describing the dispersion of the induced transmission in the non-linear experiment is composed by including into eq. (28) the value of the $\sigma_\lambda(\varepsilon)$ for the transition S_{1-0} and S_{3-2} in the form of the Lorentz contour

$$\sigma_\lambda = \sigma_0 \cdot \Delta v^2 / 4 \cdot (v - v_0) + \Delta v^2 .$$

The model curve (the red curve in fig. 31a) is close to the experimental curve. The moderate change in the relaxation time S_1 level to S_2 level from 100 ps to 150 ps leads to the well fitting of the model (blue curve) and experimental curves. The shift of the maximum of the non-linear absorption (λ=580 nm) from the maximum of the non-linear transmission (λ=575 nm) on the 150 cm^{-1} follows from the calculation. It corresponds to the shift of the luminescence maximum relative to the absorption maximum of the J-aggregates in the thin solid films.

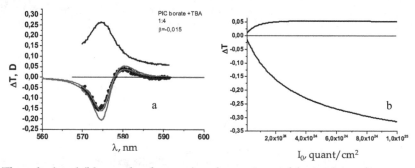

Fig. 31. The calculated (blue and red curves) and experimental (points) non-linear transmission dispersion curves for the J-aggregated PIC ($B_{10}H_{10}^{2-}$)+TBA (1:4) film (a); The calculated induced absorption (upper curve) and transmission (bottom curve) depending on the intensity laser irradiation (b)

The laser radiation intensity dependence on the non-linear transmission of the J-aggregated film sample was calculated for the obtained specified model parameters for the intensity change in the range 10^{23}-10^{25} quant/cm^2 in the maximums of the induced non-linear bleaching and darkening as shown in fig. 31b.

The experimental values of the induced darkening (ΔT=-0.05) and bleaching (ΔT=-0.2) are in good agreement with the calculated values at the experimental achieved intensities 10^6 W/cm^2 ($3 \cdot 10^{24}$ quant/cm^2). The calculated curves also coincide well with the existence of the experimentally observed saturation, as well in induced bleaching as in the induced darkening. At intensity of laser irradiation of more then $2*10^6$ W/cm^2 ($6 \cdot 10^{24}$ quant/cm^2) the irreversible burning of the dye film begins and observation of the non-linear effect becomes difficult.

Using the four-level model for J-aggregates appearing from the polarization relaxation of J-aggregate excites state is closely connected with the inhomogeneous broadening of J-aggregate absorption and luminescence spectrum, and can help us to explain the giant cubic non-linear properties of PIC J-aggregates in thin films. The big resonance values of the cubic susceptibility at the level $\chi^{(3)}$=10^{-7} esu observed in the water solutions or in the polymer matrixes at high relation polymer:dye (>10:1) are caused by the high oscillator strength of the resonance transition in two level system (Zhuravlev et al., 1992; Shelkovnikov et al., 1993). The J-aggregate in the water is surrounded by weak polarizable water molecules and the excitation-induced medium polarization has a weak influence on the change of the energy of the excited electron level of the aggregate. The non-linear response in this case is the response of the two-level system. In the solid film the J-aggregate is surrounded by highly polarizable molecules of the dye. In this case induced medium polarization leads to the deep relaxation of the energy of the excited S_1 level of J-aggregate to the relaxed state with the energy lowering to 150-170cm^{-1}. This process lasts 150 ps and in the laser pulse duration time 5 ns the effective population of the relaxed level takes place. This effect has the reflection in the linear spectrum as the inhomogeneous broadening of the absorption and luminescence contours of the J-aggregates in films. The decreasing of the population non-relaxed S_1 level leads to the essential non-linear bleaching of the J-aggregates and to the appearance of the giant non-linear susceptibility of the J-aggregates. The excitation of the populated relaxed level leads to the appearance non-linear darkening at the long wavelength slope of the J-peak in solid films. The non-linear response of the J-aggregates in the solid films is the response of the four-level system and this leads to the increasing cubic susceptibility by two orders of magnitude. The additional reason of the non-linear response increasing in solid film is the increasing of the local field factor ($F_a = \dfrac{\left(n^2 + 2\right)^2}{9n}$) due to surrounding of the J-aggregate by highly polarizable molecules of the dye. The F_a is increased by 1.65 times at the transition from dye in the solvents to the dye in the film. The cubic susceptibility depends on the four degrees of the local field factor $\chi^{(3)} = \gamma \cdot F_a^4 \cdot C$ (where C is the concentration, γ is the own polarizability of the molecule). This gives rise to the non-linear response up to seven times more.

2.11 The quantum chemical calculation of monomer and dimer PIC

The calculation of the charge distribution in the ground and excited state in the PIC2-2 molecule was carried out using the semi-empirical quantum chemical AM1 method. The charge distribution in the ground state of the cation PIC is shown in part 8 of this chapter. It coincides with the charge distribution calculated using the theory of the functional density method used in Guo et al. (2002) on the qualitative level. The charge redistribution on the atoms of the PIC molecule takes place at the excitation. Let take the transition dipole moment of the molecule be proportional to the electron density change at the excitation. The value of the electron density change between the ground and excited state was estimated by the change of the values of the dipole moments of the bounds between the neighbour atoms of the PIC π-system. The directions of the charge redistribution in the excited state of the PIC are shown in fig. 32a for one half of the molecule.

The redistribution of the electron density in the excited state of PIC leads to the next induced dipole moments in the molecule: the projection of the induced dipole moment along the X axe M_x=1.38D, along the Y axe M_y=1.44D. That means the total dipole induced by the electromagnetic field of the light wave is oscillated at the angle ~45° in the XY plane of the molecule as shown in fig. 32b.

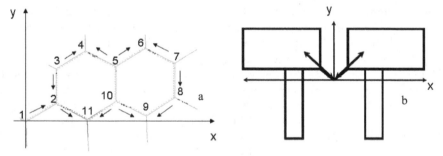

Fig. 32. The charge redistribution (a) and electron density oscillation at the PIC molecule excitation (b)

Since the directions of the projections of the induced dipoles vectors on the X axe are in opposition to each other, then only the induced dipoles vector projection directed along the Y axe is kept safe. It appears from this that the transition dipole moment is polarized along the short axis of PIC molecule.

The approximation of the dipole-dipole interaction between excited molecules in J-aggregate (Kuhn, H., Kuhn, C. 1996) does not consider the orbital overlap of the molecules. On the short distance between molecules in the aggregate, about 3-4 Å orbital overlap is essential. The existence of the orbital overlap between the neighbour dye molecules leads to the existence of the electron exchange between the orbitals and to the splitting of the electron levels of the two combined molecules. The exchange interaction has repulsive character for the system with closed shell as the dye in the ground state. This leads to the increase of energy of the two neighbour molecules in dimer. The excited state is the state with open shell and the dimer excitation takes place as as well to the high as to the low energy excited state. The energy of the splitting has the exponential decay dependence on the distance and at the distance more than 10Å is about some inverse centimetres. But at the distance between molecules in dimer 3-4Å the energy of the splitting is hundreds of the inverse centimetres.

The splitting energy induced by exchanging orbital interaction is calculated via the resonance integral (β) distance dependence. The lowering of the energy of the transition is equal β and the value of the splitting is 2β. The excited state of the long wavelength transition of the PIC is $\pi-\pi*$ excited state and it is reasonable to estimate the energy of the exchange interaction between molecules in dimer for the π–orbitals. For the estimation in the first approximation of the β value without application for the calculation of the advanced quantum-chemical methods it is possible to use the distance dependence of the resonance integral $\beta_{res}(r)$ given via calculation of the empiric integrals in the basis of the Slater-type orbitals for the π–orbitals of the conjugated double bounds in molecule that was cited in Warshell (1977).

$$\beta_{res}(r) := \left[2.438 \cdot e^{-2.035(r-1.397)} \right] \cdot [1 + 0.405 \cdot (r - 1.397)] \cdot 8065 \, (cm^{-1}) \qquad (29)$$

The calculated value β_{res} for the distance 3,4-3,6 Å is 600-420 cm^{-1}. The estimated value β_{res} gives an image of the upper boundary of the lowering of the energy for two interacting $\pi-\pi$ orbitals because it was set for the orbitals in molecule. In the some of publication the short-range electron-exchange or electron/hole charge transfer between molecules was included in the calculation for the considering of the excited state of aromatic dimers as the exciton state, (Tretiak, 2000). It was shown that intermolecular electron exchange coherent interaction leads to a crucial red shift in the dimer spectrum and completely invalidates the simple Frenkel exciton model.

The splitting and oscillator strength of the PIC-dimer depend on the configuration of the interacting orbitals and thus the arrangement geometry of the dyes in the dimer. The energetic scheme of the electron levels splitting in the dimer is shown in fig. 33. The energy of singlet transitions in absorption spectrum of PIC in monomer form was calculated using the semi-empirical quantum chemical method ZINDO/S with preliminary geometry optimization utilizing the semi-empirical quantum chemical method AM1. The calculated PIC monomer structure in the model of the overlapping spheres is shown in fig. 34. The calculated long wavelength singlet transition at 509 nm with oscillator strength 1.3 is in accordance with the absorption of molecule PIC in a vacuum (calculation from the equation 24).

The calculations of energy of transitions in PIC dimers on the basis of the supramolecular approach using the semi-empirical quantum chemical method ZINDO/S with dimer geometry optimization utilizing the molecular mechanic method MM+ shows that the two-level splitting of the energy of excited state of dimer and magnitude of splitting depends on the angle between PIC molecules. It is significant that in the previous PIC dimer calculations (Kuhn & Kuhn, 1996; Burshtein et al., 1997) the linear shift of the molecules was considered. Here we consider the model of the molecules rotation in the PIC dimer. The calculation of the energy of the dimer by MM+ method was accompanied by the optimization anion location at each step of the rotation equal to 5° of the PIC1 molecule in ZX plane relative to the PIC2 molecule as shown in fig. 35. After the whole cycle of the rotation by 90°, the distance L along the Y axe was changed and the cycle was repeated.

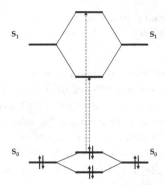

Fig. 33. The energetic scheme of the dimer levels splitting

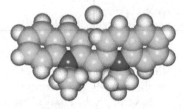

Fig. 34. The overlapping sphere model of the PIC iodide

The dimer energy distance (L) dependence for the three angles PIC molecules rotation perpendicular, parallel and at the angle of rotation 20° is shown in fig. 36. One can see that parallel molecules' orientation shows the weak energy distance dependence at the molecules' rapprochement to the 9.5Å. In further, the high energy growth in dimer takes place. The reciprocal rotation of the molecules on the 20° degree leads to clear energy distance dependence with the minimum at 7.7Å. The close energy minimum at 7.8Å has the dimer with the perpendicular molecular orientation.

The energy and oscillator strength of the PIC dimer at the rotation of the molecules divided on distance 7.8 Å were calculated using the ZINDO/S method, including 3 orbitals in configuration interactions (CI). The results of the calculation for two first allowed singlet transitions are shown in fig. 37.

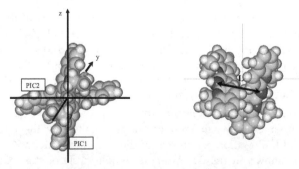

Fig. 35. The two profiles of the PIC molecules orientation in the dimer

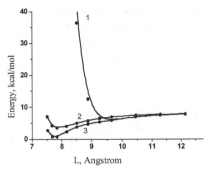

Fig. 36. The energy of the dimer molecular distance dependence for the three angles of the molecules rotation in ZX plane: 0° (1), 90° (2), 20° (3)

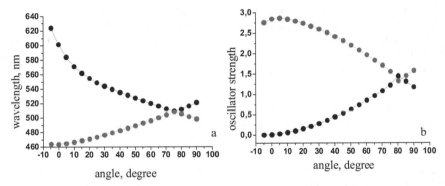

Fig. 37. The values of the wavelengths (a) and ascillator strength (b) for the two allowed singlet transitions in the PIC dimer depending on the rotation angle in ZX plane

The rotation of the dyes in dimer relative to axes connecting the centres of molecules leads to the splitting of the electronic levels giving the rise of the intensity of the short wavelength transition at parallel molecules orientation. That situation corresponds with appearance of the spectral absorption band of H-aggregates of PIC. At the angle between molecules of 75 degrees, which is close to a perpendicular orientation of the dye molecules, the energy level splitting becomes close to zero leading to the degeneration of level splitting with formation of a high intensity absorption batochromic peak with doubled oscillator strength of transition. The red shift of this dimer state in vacuum takes place to 670 cm⁻¹ if we include in the CI calculation the enlarged number of orbitals (for an example 15) to consider the more full account of electron exchange in the dimer excited state. From our point of view that situation corresponds to the formation absorption peak of J-aggregates, which has the additional red shift due to the surrounding of a high polarization medium in a thin solid film of dye.

The reason for the rotation at the dimerization of the PIC molecules is the influence of the two forces: the Coulomb interaction of the two cations and anions which try to bring the particles together and repulsive exchange interaction of the molecules' in the ground state . The overlap integral of the external π-electron orbitals of the carbon atoms and σ-electron orbitals of the hydrogen atoms is decreasing at the rotation of the molecules to the perpendicular position of the molecules regarding each other. That leads to the decreasing of the repulsive part of the molecular interaction and the molecules are brought together until the new equilibrium between the forces is established.

3. Conclusion

The medium with the high polarizability compared with the polarizability of the dye in the solutions arises as a result of the dye thin solid film formation. This creates the condition for the appearance, as well the high negative value of the non-linear cubic susceptibility in resonance as the change of the sign on the dispersion curve of the cubic susceptibility and appearance the darkening on the long wavelength slope of the J-peak absorption. In the medium with high polarizability the considering of the four-level system for the J-aggregates excited state description is more appropriate compared with the two-level system typically considered for the J-aggregates. The depopulation of J-aggregate Frank-Condon excited state due to the appearance of lowered relaxed state leads to giant non-linear bleaching of J-aggregate peak and the excitation of the relaxed state leads to the non-linear absorption appearance. The value of non-linear response strongly depends on the relaxation time. The decrease of the relaxation time leads to the growth of non-linear response.

The increase of the medium polarizability enhances the non-linear response and is reflected in the increasing of the inhomogeneous broadening in the absorption and luminescence of the J-aggregates' spectra. The formation of the medium with high polarizability is clear seen from the spectral shift of the dye absorption during spin-coating and helps to explain the J-peak growth at the solid film formation. The inhomogeneous broadening in the J-peak absorption has the connection with aggregate thermal decay energy activation distribution and spectral inequivalent in the kinetics of the thermal decay of the J-aggregate absorption counter. The medium polarization influence on the optical and non-linear optical properties is important for the aggregated or supramolecular state of different dyes and nanostructures in the condensed phase.

The number of facts allow us to consider the properties of the pseudoisocyanine J-aggregates in the thin solid films as the properties of the strongly coupled dimers with inhomogeneous broadened spectral contour: the existence of the isobestic point at the J-aggregates' thermal conversion to monomer dye in the thin solid films or in the polymer films, the absence of the any hypsochromic spectral shift at the transition of the J-aggregate to monomer in the thin films at the addition of the octadecylquinolinium iodide, the high stabilization of the J-aggregate at the addition of the dipole closo-hydrodecaborate anion, the batochromic spectral shift of the J-peak at heating of the dye thin film, the narrowing of the J-peak in the thin solid films in the presence of organic cations and others.

The practical use of the obtained results on the non-linear properties is in the possible application of cyanine dyes' J-aggregates as the elements for teraherz demultiplexing of light signals and in schemes of ultra-short laser pulse stabilization (Plekhanov et al., 2004). The radiation stability of the J-aggregates is at a level of 0.5-1 MW/cm^2. This is useful to increase the radiation stability of the J-aggregates by 5-10 times. For safer J-aggregated films, application is enough to increase the thermal stability to 100-150°C. However, the search for new highly polarized aggregates of dyes with high absorption coefficients in the visible and infra-red spectral range that possess high thermal and photochemical stability remains a basic and practical task.

4. Acknowledgment

This study was supported by the Russian Foundation for Basic Research (project no. 02-03-33336), the programme "Fundamental Problems of Physics and Chemistry of Nanosystems and Nanomaterials" of the Presidium of the Russian Academy of Sciences (grant no. 8-2) and the programme of interdisciplinary integrated investigations of the Siberian Division of the Russian Academy of Sciences (project no. 84).

5. References

Arzhantsev, S., Ito, N., Heitz, M., Maroncelli, M. (2003). Solvation dynamics of coumarin 153 in several classes of ionic liquids: cation dependence of the ultrafast component. *Chem. Phys. Lett.* Vol. 381. pp. 278-286.

Bakalis, L. D. and Knoester, J. (2000). Linear absorption as a tool to measure the exciton delocalization length in molecular assemblies. *J. Lumin.* Vol. 87-89. pp.67-70.

Bakhshiev, N. G. Libov, V. S. Mazurenko, Yu. T. Amelichev, V. A. Saidov, G. V. and Gorodynskii, V. A. (1989). Solvation Chemistry: Problems and Methods [in Russian], Leningr. Gos. Univ., Leningrad (Rus.)

Blankenship, R.E., Olson, J.M., Miller, M. (1995). Antenna complexes from green photosynthetic bacteria, In: *Anoxygenic photosynthetic bacteria.* R.E. Blankenship, M.T. Madigan, and C.E. Bauer (eds.), Kluwer Academic Publish., Dordrecht, pp. 399-435.

Bogdanov, V.L., Viktorova, E N., Kulya, S V., and Spiro, A.S. (1991). Nonlinear cubic susceptibility and dephasing of exciton transitions in molecular aggregates, *JETP Letters.* Vol. 53, pp.105-108.

Burshtein, K. Ya., Bagaturyanz, A.A., Alfimov M.V. (1997). Computer modeling of the absorption line of the J-aggregates. *Izv. AN. ser. Khim* Vol.1. pp.67-69 (Rus.)

Cerasimova, T.N., Orlova, N. A., Shelkovnikov, V.V., Ivanova, Z. M., Markov, R. V., Plekhanov, A,I., Polyanskaya, T. M., Volkov, V. V. (2000). The Structure of Pseudoisocyanine Decahydro-*closo*-decaborate and Its Nonlinear Optical Properties in Thin Films. *Chemistry for Sustainable Development.* Vol.8, pp.109-114.

Coates, E. (1969). Aggregation of dyes in aqueous solution. *JSDS.* pp. 355-368.

Daltrozzo, E., Scheibe, G., Geschwind, K., Haimerl, F. (1974). On the structure of J-aggregates of pseudoisocyanine *Photogr. Sci. Eng.* Vol. 18. № 4. pp. 441-449.],

Das, K., Sarkar, N., Das, S., Datta, A., Bhattacharya, K. (1996). Solvation dynamics in solid host. Coumarin 480 in zeolite 13X. *Chem. Phys. Lett.* Vol. 249. pp. 323-328.

Ferrante, C., Rau, J., Deeg, F.W., Brauchle, C. (1998). Solvatation dynamics of ionic dyes in the isotropic phase of liquid crystals. *J. Luminesc.* Vol. 76-77. pp. 64-67.

Furuki, M., Tian, M., Sato, Y., Pu L.S. (2000). Terahertz demultiplexing by a single short time-to-space conversion using a film of squaryliun dye J-aggregates. *Appl. Phys. Lett.* Vol.77. pp.472-474.

Ghasemi, J.B., Mandoumi, N.A. (2008). New algorithm for the characterization of thermodynamics of monomer-dimer process of dye stuffs by photometric temperature titration. *Acta Chim. Slov.* Vol. 55. pp. 377-384.

Glaeske H., Malyshev V.A., Feller K.-H. (2001). Mirrorless optical bistability of an ultrathin glassy film built up of oriented J-aggregates: Effects of two-exciton states and exciton-exciton annihilation, *J. Chem. Phys.* Vol.114. pp.1966-1969

Guo, Ch., Aydin, M., Zyu, H-R., Akins, D.L. (2002). Density functional theory used in structure determinations and Raman band assignments for pseudoisocyanine and its aggregate. *J. Phys. Chem. B.* Vol.106. pp. 5447-5454

Herz, A.N. (1974). Dye-Dye interactions of cyanines in solution and at silver bromide surfaces. *Photogr. Sci. Engineering.* Vol. 18. №3. pp. 323-335

Jelley, E., (1936). Spectral absorption and fluorescence of dyes in the molecular state. *Nature,* Vol.138, pp. 1009-1010.,

Katrich, G.S., Kemnitz, K., Malyukin, Yu.V., Ratner, A.M. (2000). Distinctive features of exciton self-trapping in quasi-one-dimensional molecular chains (J-aggregates). *J. Luminesc.* Vol. 90. pp. 55-71

Kobayashi, T., ed. (1996). *J-aggregates,* World Scientific Publish. Co Pte. Ltd., Singapore.

Krasnov, K.S., (1984). *Molecules and chemistry bond,* High school. Moscow. (Rus.)

Kuhn, H., Kuhn, C. (1996). Chromophore coupling effects. In: *J-Aggregates.* T. Kobayashi (ed.)- Singapore: World scientific publishing Co. Pte. Ltd., - 228 p.

Lakowics, J.R. (1983). *Principles of fluorescence spectroscopy.* Plenum press. New York and London

Levshin, L.V., Salecky, A.M. (1989). *Luminescence measurement.* MGY. Moscow. (Rus.)

Mal'tsev E.I., Lypenko D.A., Shapiro B.I., and Brusentseva M.A. (1999). Electroluminescence of polymer/J-aggregate composites. *Appl. Phys. Lett.* Vol. 75, pp.1896-1898.

Malyshev, V.A. (1993). Localization length of one-dimensional exciton and low-temperature behaviour of radiative lifetime of J-aggregated dye solutions. *J. Luminesc.* Vol. 55. pp. 225-230.

Mandal, D., Sen, S., Bhattacharya, K., Tahara, T. (2002). Femtosecond study of solvation dynamics of DCM in micelles. *Chem. Phys. Lett.* Vol. 359. pp. 77-82.

Markov, R. V., Chubakov, P. A., Plekhanov, A. I., Ivanova, Z. M., Orlova, N. A., Gerasimova, T. N., Shelkovnikov, V. V., and Knoester, J. (2000). Optical and nonlinear optical properties of low-dimensional aggregates of amphyphilic cyanine dyes. *Nonlinear Opt.,* Vol.25, pp.365–371

Markov, R.V., Plekhanov, A.I., Rautian, S.G., Orlova, N.A., Shelkovnikov, V.V., Volkov, V.V. (1998). Nonlinear optical properties of two types of PIC J-aggregates in thin films. *Proc. SPIE.* Vol. 3347. pp. 176-183. (a)

Markov, R.V., Plekhanov, A.I., Rautian, S.G., Orlova, N.A., Shelkovnikov, V.V., Volkov, V.V. (1998). Nonlinear optical properties of the two types pseudoisocyanine J-aggregates in thin films. *Jurnal Nauchn. and Prikl. Fotogr.* Vol. 43. pp.41-47. (Rus.) (b)

Markov, R.V., Plekhanov, A.I., Rautian, S.G., Safonov, V.P., Orlova, N.A., Shelkovnikov, V.V., Volkov, V.V. (1998). Dispersion of cubic susceptibility of thin films of pseudoisocyanine J-aggregates as measured by longitudinal scanning, *Optics and Spectroscopy.* Vol. 85. pp.588-594. (c)

McDermott, G., Prince, S.M., Freer, A.A. et al., (1995). Crystal structure of an integral membrane light-harvesting complex from photosynthetic bacteria, *Nature.* Vol.374. pp. 517-521.

Minoshima, K., Taiji, M., Misawa K. and Kobayashi, T. (1994). Femtosecond nonlinear optical dynamics on excitons in J-aggregates. *Chem. Phys. Lett.* Vol.218. pp.67–72

Moll, J., Daehne, S., Durrant, J. R. and Wiersma, D. A. (1995). Optical dynamics of excitons in J-aggregates of carbocyanine dye. *J. Chem. Phys.* Vol.102. No. 16. pp.6362-6370

Nygren, J., Andrade, J. M., Kubista, M. (1996). Characterization of a single sample by combining thermodynamic and spectroscopic information in spectral analysis. *Anal. Chem.* Vol. 68. pp. 1706-1710.

Orlova, N.A., Kolchina, E.F., Zhuravlev, F.A., Shakirov, M.M., Gerasimova, T.N., Shelkovnikov, V.V.(2002). Synthesis of 2,2'-quinocyanines with long alkyl groups. *Chemistry Heterocyclic Compounds.* № 10, pp. 1399-1407. (Rus.)

Orlova, N.A., Zhuravlev, F.A., Shelkovnikov, V.V., Gerasimova, T.N. (1995). Synthesis of pseudoisocyanines with nonsaturated groups in position 1. *Izv. AN, ser, khim.* № 6, pp. 1122-1124 (Rus.)

Pilling, R.L., Hawthorne, M.F. (1964). The boron-11 nuclear magnetic resonance spectrum of $B_{20}H_{18}^{-2}$ at 60 Mc./sec.*J. Amer. Chem. Soc.* p Vol. 86. pp. 3568-3569.,

Plekhanov A.I., Rautian S.G., Safonov V.P., (1995*). Optics and Spectroscopy* Vol.78, 1, p.92. .(Rus.)

Plekhanov, A.I., Kuch'yanov, A.S., Markov, R.V., Simanchuk, A.E., Avdeeva, V.I., Shapiro, B.I., Shelkovnikov, V.V. (2004). Passive mode locking of a Nd3+:YAG laser with a saturable absorber in the form of thin film of J-aggregates. *J. Nonlinear Org. and Polymer Materials.* Vol. 9. № 3. pp. 503–511.

Plekhanov, A.I., Orlova, N.A., Shelkovnikov, V.V., Markov, R.V., Rautian, S.G., Volkov, V.V. (1998). Third-order non-linearity optical properties of the film of the cyanine dye with borate anion. *Proc. SPIE.* Vol. 3473. pp. 100-107. (a)

Plekhanov, A.I., Rautian, S.G., Safonov, V.P., Chubakov, P.A., Orlova, N.A., Shelkovnikov, V.V (1998). Dispersion of the real and imaginary parts of cubic susceptibility in submicron films of pseudoisocyanine J-aggregates. *Proc. SPIE.* Vol. 3485. pp. 418-424.(b)

Plekhanov, A.I., Markov, R.V., Rautian, S.G., Orlova, N.A., Shelkovnikov, V.V., Volkov, V.V (1998). Third-order nonlinearity optical properties of the films of cyanine dye with borate anion. *Proc. SPIE "Third-Order Nonlinear Optical Materials".* Vol. 3473. pp. 20-31.(c)

Renge, I., Wild, U.P. (1997). Solvent, temperature, and excitonic effects in the optical spectra of pseudoisocyanine monomer and J-aggregates. *J. Phys. Chem. A.* Vol. 01. pp. 7977-7988.

Saha, S., Mandal, P.K., Samanta, A. (2004). Solvation dynamics of Nile Red in a room temperature ionic liquid usingstreak camera. *Chem. Phys.* Vol. 6. pp. 3106-3110.

Sato, T., Yonezawa, Y., Hada, H. (1989). Preparation and luminescence properties of J-aggregates of cyanine dyes at the phospholipid vesicle surface. *J. Phys. Chem.* Vol. 93. pp. 14-16.

Scheibe, G., (1936). Variability of the absorption spectra of some sensitizing dyes and its cause. *Angew. Chem,* Vol.49, p. 563.

Shapiro, B.I. (1994) Aggregates of cyanine dyes: photographic problems, *Russian Chemical Reviews.* Vol.63. pp.231-241.

Shelkovnikov, V. V., Ivanova, Z. M., Orlova, N. A., Volkov, V. V., Drozdova, M. K., Myakishev, K. G., Plekhanov, A. I.(2004). Optical Properties of Solid Pseudoisocyanine Films Doped with Cluster Derivatives of Boron Hydrides. *Optics and Spectroscopy.* Vol. 96, pp.824-833

Shelkovnikov, V.V., Plekhanov, A.I., Safonov, V.P., Zhuravlev, F.A. (1993) Nonlinear optical properties of the assembliesof organic molecules and fractal metal clusters. *Zhurnal struct. chem.* Vol.34. pp.90-105 (Rus.)

Shelkovnikov,V. V., Ivanova, Z. M., Orlova, N. A., Gerasimova, T. N., Plekhanov, A. I. (2002). Formation and properties of long alkyl substituted pseudoisocyanines J-aggregates in thin films. *Optics and Spectroscopy.* Vol.92. pp. 958–966

Struganova, I. (2000) Dynamics of formation of 1,1'-diethyl-2,2'-cyanide iodide J-aggregates in solution. *J. Phys. Chem. A.* Vol. 104. № 43. pp. 9670-9674.

Sundstrom, V., Gillbro, T., Gadonas, R.A., Piskarskas, A. (1988). Annihilation of singlet excitons in J-aggregates of pseudoisocyanine (PIC) studied by pico- and subpicosecond spectroscopy. *J. Chem. Phys.* Vol. 89. № 5. pp. 2754-2762.

Suppan, P. (1990). Solvatochromic shifts - the influence of the medium on the energy of electronic states. *J. Photochem. Photobiol. A: Chemistry.* Vol. 50. pp.293-330.

Tani, I.., Liu-Yi, Sasaki, F., Kobayashi, S., Nakatsuka, H. (1996). Persistent spectral hole-burning of pseudoisocyanine bromide J-aggregates. *J. Luminesc.* № 66-67. pp.157-163.

Tani, T. (1996) J-aggregates in spectral sensitization of photographic materials, In: *J-aggregates.* T. Kobayashi (ed.). World Scientific Publish. Co Pte. Ltd., Singapore. pp. 209-228.,

Tikhonov, E.A., Shpak, M.T. (1977). *Nonlinear optical phenomena in organic compounds.* Naukova dumka. Kiev.

Tretiak, S., Saxena, A., Martin, R. L., and Bishop, A. R. (2000). Interchain Electronic Excitations in Poly(phenylenevinylene) (PPV) Aggregates. *J. Phys. Chem. B,* Vol.104, pp.7029-7037

Trosken, B., Willig, F., Spitle, R. M. (1995) The primary steps in photography: excited J-aggregates on AgBr microcrystals, *Advanced Materials,* Vol. 7, pp. 448-450. ,

Vacha, M., Furuki, M., Tani, T. (1998). Origin of the long wavelength fluorescence band in some preparations of J-aggregates low-temperature fluorescence and hole burning study. *J. Phys. Chem. B.* Vol. 102. pp. 1916-1919.,

Verezchagin A.N. (1980). *Polarisability of molecules.* Nauka. Moskow. (Rus.)

Wang Y. (1991) Resonant third-order optical nonlinearity of molecular aggregates with low-dimensional excitons, *Journal of the Optical Society of America B.* Vol.8, pp.981-990.,

Warshell A. (1977). The self-consistent force field method and quantum chemical generalization, In: *Semiempirical methods of electronic structure calculation.* Segal. G.A. (ed.). Plenum press. New York and London

Wendlandt W.W. (1974). *Thermal Methods of Analysis.* John Wiley & Sons, Inc. New York.

Wurthner, F., Kaiser, Th.E., Saha-Moller, Ch.R. (2011). J-Aggregates: From Serendipitous Discovery to Supramolecular Engineering of Functional Dye Materials. *Angew. Chem. Int. Ed.* Vol.50. pp. 3376 – 3410.

Zhuravlev, F.A., Orlova, N.A., Shelkovnikov, V.V., Plekhanov, A.I., Rautian, S.G., Safonov, V.P. (1992)Giant non-linear susceptibility of the thin films with complexes molecular aggregate – metal cluster. *Pis'ma JETF.* Vol.56. pp.264-267 (Rus.)

7

Flow-Injection Spectrophotometric Analysis of Iron (II), Iron (III) and Total Iron

Ibrahim Isildak

Yildiz Technical University, Faculty of Chemical and Metallurgical Engineering,
Bioengineering Department, İstanbul,
Turkey

1. Introduction

Determination of iron in analytical chemistry has become a routine procedure because of its importance in our life. Various chemical forms of iron can be found in natural waters depending on geological area and chemical components present in the environment. The main source of iron in natural waters is from the weathering and leaching of rocks and soils (Dojlido & Best, 1993). Also, metallic iron and its compounds are used in various industrial processes and may enter natural waters through the discharge of wastes. Iron(II) is normally less present in river water (Sangi et al., 2004) and iron (III) can precipitate rapidly by the formation of hydrous iron oxide and hydroxides, which they can absorb other trace metals. Thus, iron ion controls the mobility, bioavailability and toxicity of other trace metals in the natural water system (Wirat, 2008; Lunvongsa et al., 2006). Amounts of iron are widely present in tap, pond, well and underground water, and this metallic ion is essential for biological systems (Ohno et al., 2004; Kawakubo et al., 2004).

As iron is one of the most frequently determined analyte in environmental (water, soil and sediment) samples, many spectrophotometric and/or flow-injection spectrophotometric methods have been developed for iron determination. When trace levels of the iron are concerned, the detection methods applicable are reduced (Tarafder et al., 2005; Weeks & Bruland, 2002; Giokas et al., 2002; Themelis et al., 2001; Bagheri et al., 2000; Pascual-Reguera et al., 1997; Teshima et al., 1996; Tesfaldet et al., 2004; Udnan et al., 2004; Pojanagaroon et al., 2002; van Staden & Kluever, 1998; Asan et al., 2003, 2008; Andac at al., 2009). Flow-injection analysis, as a rapid and precise technique, has found wide application in the determination of iron in several sample matrices (Bowie A.R., et al. 1998; Hirata S., et al. 1999; Qin W., et al. 1998; Kass M., et al. 2002; Saitoh K., et al. 1998; Weeks D.A., et al. 2002; Giokas D.L., et al. 2002; Themelis D.G., et al. 2001; Bagheri H., et al. 2000; Molina-Diaz A., et al. 1998; Teshima N., et al. 1996).

Highly sensitive, selective and rapid flow-injection spectrophotometric analysis (FIA) methods for the determination of iron (II), iron (III) and total iron will be defined under proposed chapter of the book. The methods were based on the reactions of iron (II) and iron (III) with different complexing agents in different carrier solutions in FIA (Asan A. et al., 2010; Andac M. et al., 2009; Asan A. et al., 2008). Several parameters acting on the determination of iron (II) and iron (III) were examined. The developed methods have been successfully applied to the determination of iron (II), iron (III) and total iron in water and ore samples. The methods were also verified by applying certified reference materials.

2. A very sensitive flow-injection spectrophotometric determination method for iron(II) and total iron using 2', 3, 4', 5, 7-pentahydroxyflavone

Spectrophotometric detection based on the measurement of the absorbance at a characteristic wavelength of complex formed between a chelating agent and iron has been mainly applied (Kass M. and Ivaska A. 2002; Saitoh K., et al. 1998; Weeks D.A. and Bruland K.W. 2002; Giokas D.L., et al. 2002; Themelis D.G., et al. 2001; Bagheri H., et al. 2000; Molina-Diaz A., et al. 1998; Teshima N., et al. 1996; Tesfaldet Z.O., et al. 200); Udnan Y., et al. 2004; Morelli B., et al. 1983; Pojanagaroon T., et al. 2002; van Staden J.F. and Kluever L.G. 1998). A number of other chelating agents that have been reported for the spectrophotometric and/or flow-injection spectrophotometric determination of iron (III) and total iron include 2-thiobarbituric acid (Morelli B., et al. 1983), norfloxacin (Pojanagaroon T., et al. 2002), tiron (Mulaudzi L.V., et al. 2002; Van Staden J.F. and Kluever L.G. 1998), tetracycline (Ahmed M.J. and Roy U.K. 2009) and chlortetracycline (Sultan S.M. and Suliman F. 1992). Flow-injection spectrophotometric methods based on above chelating agents are not either selective, or a masking agent should be used (Wirat R., 2008). However, highly selective, simple and economical methods are still required for the routine determination of iron (II) in different sample matrices. An ultra-sensitive and highly selective, rapid flow-injection spectrophotometric method for the determination of iron (II) and total iron has been proposed. The method was based on the reaction between iron (II) and 2', 3, 4', 5, 7-pentahydroxyflavone (Morin) in slightly acidic solution (pH:4.50) with a strong absorption at 415 nm. The chemical structure of Morin is shown Fig. 1. The reagent itself is sparingly soluble in water and does not absorb in the visible region of the spectrum, therefore, might be well suited for flow-injection analysis of iron (II) and total iron. The method has been successfully applied to the determination of iron (II) and total iron in water samples and ore samples.

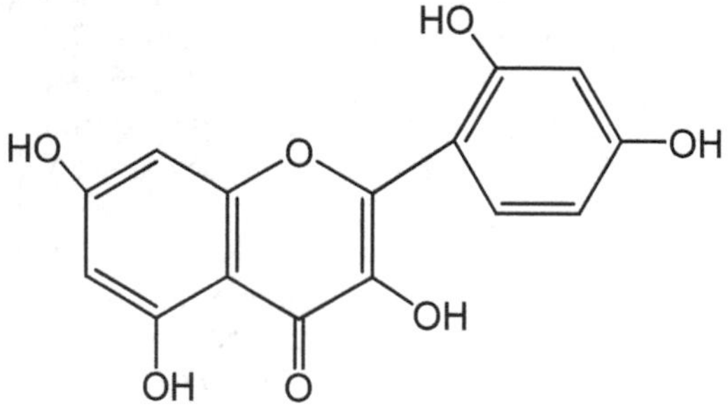

Fig. 1. The chemical structure of 2', 3', 4', 5', 7-pentahydroxyflavone (Morin)

2.1 Experimental

2.1.1 Reagent and standards

All chemicals used were of analytical reagent grade or the highest purity available. Doubly distilled deionized water was used throughout the study. Glass vessels were cleaned by soaking in acidified solutions of $KMnO_4$ or $K_2Cr_2O_7$ followed by washing with concentrated HNO_3, and were rinsed several times with high-purity deionized water. Stock solutions and environmental water samples (1000 mL each) were kept in polypropylene bottles containing 1 mL of concentrated hydrochloric acid. Standard iron (II) and iron (III) stock solutions were prepared by solving 278.02 mg of iron (II) and 489.96 mg of iron (III) sulphate (Merck) in 0.01 M 100 mL hydrochloric acid to give 0.01 M stock solution of iron (II) and iron (III). Iron (II) and iron (III) working standard solutions were prepared daily by suitable dilution of stock solutions with double deionized water. Standard reference material consisting of 0.085 % Fe (Zn/Al/Cu 43XZ3F) was provided from MBH Analytical Ltd. (UK). Hydrogen peroxide solution 30 % (v/v) used was from Merck.

A stock solution of Morin (5×10^{-3}M) was prepared by dissolving requisite amount of Morin (BDH Chemicals) in 100 mL of ethanol:water (4:96 v/v) because of it's low solubility in water only. For spectrophotometric study, morin complex solutions of various metals were prepared by mixing 1 mL of 1×10^{-4} M standard solution of each metal in double deionised water with a suitable volume of 1×10^{-4} M Morin solution. All stock solutions were stored in polyethylene containers. All polyethylene containers and glassware used for aqueous solutions containing metallic cations were cleaned with (1+1) nitric acid while the rest were cleaned with 3 % Decon 90, all were rinsed with deionized water before use. The working standard solutions were prepared by appropriate dilution immediately before use. All solutions were degassed before use using a sonicator (LC 30). Reagent carrier solution was composed of Morin in 0.1 M HAc/Ac- buffer (pH:4.50) solution consisting of metanol 4 %.

2.1.2 Apparatus

UV-Visible spectra of metal-AcSHA complexes were taken with a Unicam spectrophotometer (GBC Cintra 20, Australia). A Jenway 3040 Model digital pH-meter was used for the pH measurements.

In the FIA system, peristaltic pump (ISMATEC; IPC, Switzerland) 0.50 mm i.d. PFTE tubing was used to propel the samples and reagent solutions. Samples were injected into the carrier stream by a 7125 model stainless steel high pressure Rheodyne injection valve provided with a 20 μL loop. The absorbance of the coloured complex formed (λ_{max} 415 nm) was measured with a UV-Visible spectrophotometer equipped with a flow-through micro cell (Spectra SYSTEM UV 3000 HR, Thermo Separation Products, USA), and connected to a computer incorporated with a PC1000 software programme.

A UNICAM 929 model (Shimadzu AA-68006) flame atomic absorption spectrophotometer with deuterium-lamp background correction was used for the determination of iron in reference to the FIA method. The measuring conditions were as follows: UNICAM hollow cathode lamp, 10 cm 1-slot burner, air-acetylene flame (fuel gas flow-rate 1.50 L/min), 0.2 nm spectral bandwidth, and 7 mm burner height. The wavelength and the lamp current of Fe was respectively 248 nm and 5 mA.

2.1.3 General procedure

The FIA system used was simple as shown schematically in Fig.2. The sample solution was introduced into the reagent carrier solution by the Rhodyne injection valve. The complex (λ_{max}=415 nm) was formed on passage of the reagent and iron (II) ion solution through the mixing coil. A PTFE tubing (50 cm long) was attached before the flow-through detection cell as a mixing coil. The absorbance of the coloured complex was selectively monitored in the flow-through spectrophotometric cell at 415 nm. The transient signal was recorded as a peak, the height of which was proportional to the iron (II) concentration in the sample, and was used for all measurements. Five replicate injections per sample were made.

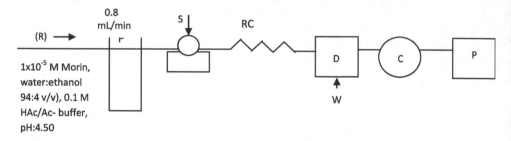

Fig. 2. Flow diagram of the FIA system used. R; reagent carrier solution (1×10^{-5} M Morin in ethanol: water (4:96 v/v) in 0.1 M HAc/Ac- buffer (pH:4.50)), P, Peristaltic pump, S; Rheodyne sample injection valve, RC; reaction coil (50 cm long, 0.5 mm i.d), D; spectrophotometric detector (λ_{max} = 415 nm), W; waste, C; computer, P; printer.

2.1.4 Sample preparation procedures

Sea, river and industrial water samples collected in Nalgene plastics were acidified by adding 1 mL of hydrochloric acid (0.1 M) per 100 mL of sample solution behind filtration over 0.45 µm Millipore Filter (Millford, MA). After filtration, 20 µL of water samples were injected directly into the FIA system for the determination of iron (II). Total iron was determined by reducing of all forms of iron to iron (II) in the procedure described (van Staden J.F. and Kluever L.G. 1998; Asan A., et al. 2003).

A 0.10 g sample of the certified metal alloy (Zn/Al/Cu 43XZ3F) was dissolved in 12 mL of concentrated $HCl+HNO_3$ (3:1 v/v) in 100 mL beaker. The mixture was heated on a hot plate nearly to dryness; 5 mL HNO_3 was added to complete dissolution and diluted to 100 mL with deionized water. The solution was filtered and transferred quantitatively to 1000 mL volumetric flask and made up to volume with deionized water. 9 mL of this solution was treated with 1 mL of sodium azide (2.5 % w/v) for iron (III) reduction. After the reduction step, 20 µL of this solution was used for the determination of total iron (van Staden J.F. and Kluever L.G. 1998).

Metal ore samples (0.10 g) were powdered ($\geq$ 500 mesh) and prepared as in the procedure described above. All analyses were performed with the least possible delay.

2.2 Results and discussion

2.2.1 Spectrophotometric studies of the Morin-metal complexes

The reaction mechanism of the present method was as reported earlier (Busev A.I., et al. 1981). Job's method of continuous variation and the molar ratio method were applied to ascertain the stoichiometric composition of the complex (MacCarthy P. and Zachary D.H.,

1986). A Fe(II)-Morin (1:2) complex was indicated by both methods. The reaction was very fast. Metal ions react with Morin in aqueous medium in the range pH: 2.0-7.0 forming coloured complexes with different stoichiometry. Absorption spectra's those correspond to solutions of 5×10^{-5} M of iron (II)-Morin complex was measured against a reagent blank and the average molar absorption coefficient of 6.82×10^4 L mol^{-1} cm^{-1} are shown in Fig. 3.

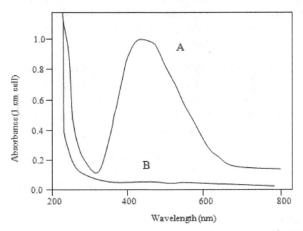

Fig. 3. Absorption spectras of iron (II)-Morin complex and Morin itself. (A) absorption spectra of iron (II)-Morin complex (5×10^{-5} M) and (B) absorption spectra of the Morin in aqueous solution.

As can be seen from the Fig. 3, the iron (II) Morin complex that has an absorbance maxima at 415 nm. At this wavelength, the Morin itself has no absorption while Morin complexes of all of the tested metal ions and the anions (not shown) exhibited a negligible absorption.

In order to develop an FIA method based on the above phenomenon, the FIA setup shown in Fig.1 was used. In the FIA system, a complex was formed with an absorption spectrum that showed a maximum at 415 nm, which was in agreement with the value obtained in the spectrophotometric study.

2.2.2 Optimisation of chemical variables and the FIA manifold

Various variables closely related to the iron determination were examined using the simple flow-injection analysis system with a fixed iron (II) concentration of 5 µg L^{-1}. The Morin concentration was varied from 1×10^{-6} M to 1×10^{-2} M. The peak height was found to increase with increasing Morin concentration up to 1×10^{-5} M and no noticeable increase was found at higher concentrations. Therefore, 1×10^{-5} M Morin was decided as colour developing component of the carrier solution.

With the concentration of the Morin fixed 1×10^{-5} M, the pH of the carrier solution was varied from 2.0 to 7.0. The interference effect of the iron (III) were found to increase with increasing pH up to 4.5 and remain constant at higher pH. Also, the peak heights were found to increase with increasing pH up to 4.0, remain constant to 4.5 and decreased slightly above that.

The pH of the reagent carrier was however adjusted to 4.5 to obtain maximum peak height and minimum iron (III) interference in the analysis. In order to proceed with the final system design, the effect of sample volume, mixing coil length and flow-rate were studied using Morin at fixed concentration of 2.5×10^{-4} M and pH 4.5.

The sample volume was varied from 5-50 µL. The peak height was decreased by decreasing sample size, and the peaks were broadened with increasing sample size due to sample zone dispersion. A sample injection volume of 20 µL was selected as a compromise between sensitivity and sample throughput rate.

The mixing coil (RC) was examined by using PTFE tubing's (0,5 mm i.d.) at different lengths ranging between 10 and 150 cm. The peak height was increased with increasing mixing coil length from 10-50 cm. The peak height was decreased for lower concentrations and broadened for higher concentrations at longer coil lengths. A mixing coil length 50 cm was decided convenient for better peak height and shape.

The flow-rate was varied from 0.2 to 2 mL min^{-1}. The peak height decreased by increasing flow-rate, probably the extent of reaction decreased. A flow-rate of 0.8 mL min^{-1} was selected as a compromise between sample throughput rate and sensitivity.

2.2.3 Calibration, accuracy and precision

The developed analytical method was validated by evaluating the linear dynamic range, precision, accurate, limit of detection (LOD) and limit of quantification (LOQ) as well as by applying the standard addition technique. Under the optimized experimental conditions, a linear calibration graph was obtained for 0.01-120 mg L^{-1} iron (II) under the optimum conditions with a regression coefficient of 0.9914. The relative standard deviation for the determination of 5 µg L^{-1} iron (II). was 0.85 % for 10 replicate injections. The limit of detection (blank signal plus three times the standard deviation of the blank) was 0.4 µg L^{-1}. The sample throughput of the proposed method was almost 60 sample h^{-1}.

2.2.4 Interference studies

The interference effects of many cations and anions on the determination of 5 µg L^{-1} iron (II) were examined. The results summarized in Table 1.

Tolerance limit (µg L^{-1})	Foreign ion
Over 50000	Cr(III), Al(III), Cd(II), Mn(II), K(I), Na(I), Ag(I), Ca(II), Mg(II), Ba(II), Hg(II), CN$^-$, NO$_3^-$, NO$_2^-$, SO$_4^{2-}$, CO$_3^{2-}$, Cl$^-$, Br$^-$, PO$_4^{3-}$, NH$_4^+$, SCN$^-$, tartrate, oxalate, citrate, thio-urea
Over 200	Fe (III)

Table 1. Effect of foreign ions on the determination of 5 µg L^{-1} of iron (II) in solution

In the table, the tolerable concentration of each diverse ion was taken as a highest concentration causing an error of ± 3 %. Most of the ions examined did not interfere with the determination of iron (II). The major interference was iron (III) at the amounts of 200 µg L^{-1}.

It is apparent from the Table 1 that the proposed method can tolerate all of the interfering species tested in satisfactory amounts and it is therefore adequately selective for the determination of Fe (II) and total iron.

2.2.5 Applications

The FIA method was applied to the determination of iron (II) and total iron in water samples and ore samples. In order to evaluate the accuracy of the proposed method, the determination of total iron in a standard reference material (Zn/Al/Cu 43XZ3F) and metal alloy sample was carried out. The analytical results obtained by the proposed method are in good agreement with the certified values as is shown in Table 2.

Sample	Total Fe[1] (%)	Certified Fe (%)
Alloy (1)	8.23(0.12)	8.58
Alloy (2)	16.15(0.16)	16.62
Std Zn/Al/Cu 43XZ3 F	0.083(0.02)	0.085

Values in parenthesis are the relative standard deviations for n=5 with confidence level of 95 %.

Table 2. Total iron content of iron alloys and standard reference material

For the application of the proposed FIA method to river and sea water samples collected from different sources were analyzed by using both calibration curve and standard addition methods. The values obtained from the calibration curve and the standard addition methods are in good agreement with each other as shown in Table 3.

Samples[1]	Iron (II)[2] (μg L-1)		Total iron (2) (μg L-1)		
	Found[3]	Found[4]	Found[3]	Found[4]	AAS
Kurtun river water	38.33(0.24)	38.55(0.12)	42.33(0.02)	42.91(0.18)	43.65(0.17)
Seaport sea water	68.84(0.32)	68.65(0.24)	85.13(0.12)	85.75(0.06)	86.12(0.12)
Baruthane sea water	47.51(0.18)	47.62(0.14)	57.24(0.04)	57.65(0.15)	58.97(0.24)
Organized industry water	78.84(0.22)	78.65(0.18)	78.13(0.14)	98.75(0.07)	99.12(0.10)

[1] Samples were collected at Samsun, Turkey.
[2] Values in parenthesis are the relative standard deviations for n=5 with confidence level of 95 %.
[3] Calibration curve method.
[4] Standard addition method.

Table 3. Determination of iron (II) and total iron in river and sea water samples

Atomic absorption measurements taken in water samples were also given for comparison in Table 3. The analytical value of total iron in water is slightly in good agreement with that obtained by the AAS method. The results obtained show that the proposed method can be applied in the determination of iron (II) and total iron content in the water samples without a pre-concentration process.

3. Flow injection spectrophotometric determination of iron (III) using diphenylamine-4-sulfonic acid sodium salt (Reproduced with permission from the paper of Asan Adem et al., 2008. Copyright of Institute of Chemistry, Slovak Academy of Sciences)

In recent years, low cost automatic and userfriendly analytical methods have become attractive for the determination of trace levels of iron in many kinds of samples (Chen et al., 2006; Pons et al., 2005a; Lunvongsa et al., 2006b). Among these, flow-injection analysis (FIA) is a well accepted technique owing to its high sample throughput, cost effective performance, versatility, flexibility, and ease of operation. Also, FIA is compatible with a wide range of detection systems (Guo & Baasner, 1993; Ensafi et al., 2004). Up to date, FIA for the determination of iron(III) has been generally combined with optical detectors (Pulido-Tofino et al., 2000; Saitoh et al., 1998). The spectrophotometric detector based on measuring the absorbance of colored complexes formed with various chromogenic reagents is one of the most frequently used detectors for the determination of iron in many kinds of samples (Yegorov et al., 1993; Yamamura and Sikes, 1966; Ampan et al., 2002; Bruno et al., 2002; Tesfaldet et al., 2004; Van Staden and Kluever, 1998; Mulaudzi et al., 2002; Reguera et al., 1997; Pojanagaron et al., 2002; Araujo et al., 1997; Asan et al., 2003; Udnan et al., 2004; Alonso et al., 1989; Müller et al., 1990; Themelis et al., 2001; Kass and Ivaska, 2002; Weeks and Bruland, 2002). A large number of flow-injection spectrophotometric methods have been developed for the determination of iron using desferal (Yegorov et al., 1993), 1,10-phenantroline (Yamamura and Sikes, 1966; Ampan et al., 2002; Bruno et al., 2002; Tesfaldet et al., 2004), tiron (Van Staden and Kluever, 1998; Mulaudzi et al., 2002), ferrozine (Reguera et al., 1997), norfloxaxin (Pojanagaron et al., 2002), thiocyanate (Araujo et al., 1997), DMF (Asan et al., 2003), and salicylate (Udnan et al., 2004) as chromogenic reagents. However, many of the proposed methods have a high limit of detection (Alonso et al., 1989; Müller et al., 1990), suffer from many interfering metal ions, such as Zn and Co (Guo and Baasner, 1993), have a short linear dynamic range (Themelis et al., 2001; Kass and Ivaska, 2002), tedious procedures (Pons et al., 2005b), or low sampling rates (Teixeira and Rocha, 2007; Lunvongsa et al., 2006a).

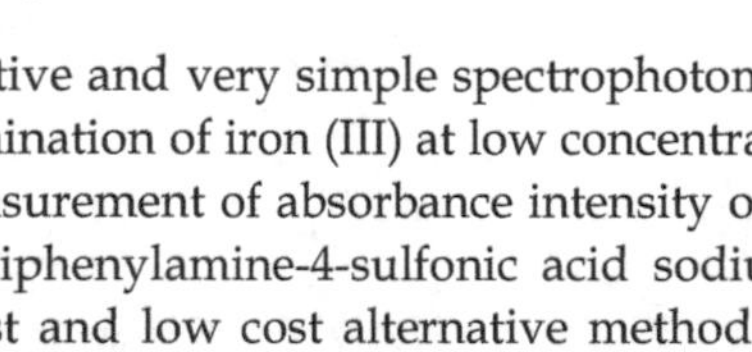

Fig. 4. Structure of the diphenylamine-4-sulfonic acid sodiumSalt

In this study, a highly sensitive and very simple spectrophotometric flow-injection analysis (FIA) method for the determination of iron (III) at low concentration levels is presented. The method is based on the measurement of absorbance intensity of the red complex at 410 nm formed by iron (III) and diphenylamine-4-sulfonic acid sodium salt (DPA-4-SA). It is a simple, highly sensitive, fast and low cost alternative method using the color developing reagent DPA-4-SA in acetate buffer at pH 5.50 and the flow-rate of 1 mL min^{-1} with the sample throughput of 60 h^{-1}. The accuracy of the method was evaluated using the standard addition method and checked by the analysis of the certified material Std Zn/Al/Cu 43 XZ3F.

3.1 Experimental

3.1.1 Reagents, chemicals, equipment

All reagents used were of analytical reagent grade and the solutions were prepared with double distilled and deionized water. The reagent diphenylamine-4- sulfonic acid sodium salt (DPA-4-SA) was provided by Merck. The chemical formula of the DPA-4-SA is shown in Fig. 4. Standard iron(III) (1 mg mL^{-1}) and iron(II) (5 mg mL^{-1}) solutions were prepared by dissolving FeCl$_3$ 6H$_2$O and FeCl$_2$ 4H$_2$O in 0.05 M nitric acid and standardized by titration with EDTA. The stock solution of DPA-4-SA (1×10$_{-2}$ M) was prepared by dissolving the diphenylamine-4-sulfonic acid sodium salt in deionized water. All stock solutions were stored in polyethylene containers. All polyethylene containers and glassware used for aqueous solutions containing metallic cations were cleaned with (1+1) nitric acid while the rest were cleaned with 3 % Decon 90, all were rinsed with deionized water before use. The working standard solutions were prepared by appropriate dilution immediately before use. Interference studies were carried out using chloride or nitrate salts of the metal cations, and sodium or potassium salts of anions. All solutions were degassed before use using a sonicator (LC 30). A certified metal alloy sample consisting of 0.085 % Fe (Zn/Al/Cu 43XZ3F) was provided by MBH Analytical Ltd. (UK).

The pH measurements were carried out using a Jenway 3040 Model digital pH-meter consisting of a contained glass pH electrode. UV-Visible spectra of the DPA-4-SA reagent and metal-DPA-4-SA complexes were taken with a Unicam spectrophotometer (GBC Cintra 20, Australia). A peristaltic pump (ISMATEC; IPC, Switzerland) was used to propel the samples and reagent solutions. Samples were injected into the carrier stream by a 7125 model stainless steel high-pressure Rheodyne injection valve provided with a 20 µL injection loop. Absorbance of the colored complex formed in the flow system was measured using a UV-visible spectrophotometer equipped with a flowthrough micro cell (Spectra SYSTEM UV 3000 HR, Thermo Separation Products, USA), and connected to a computer (IPX Spectra SYSTEM SN 4000) incorporated with a PC 1000 software program. The reaction coil was made of PTFE tubing (1 m, 0.5 mm, i.d.). A UNICAM 929 model (Shimadzu AA-68006) flame atomic absorption spectrophotometer with deuterium- lamp background correction was used for the determination of iron in reference to the FIA method. The measuring conditions were as follows: UNICAM hollow cathode lamp, 10 cm 1-slot burner, airacetylene flame (fuel gas flow-rate 1.50 L min^{-1}), 0.2 nm spectral bandwidth, and 7 mm burner height. The wavelength and the lamp current of iron were 248 nm and 5 mA, respectively.

The manifold of the flow-injection system was similar to that proposed in our previous study (Asan et al., 2003). The peristaltic pump was used for propelling the reagent carrier solution at a flow-rate of 1 mL min^{-1}. Samples were injected into the reagent carrier solution, soon load the reaction coil. The reaction zone containing the complex was moving towards the flow-through spectrophotometric detector cell in which the presence of iron(III)-DPA-4-SA complex was selectively monitored, and the absorbance of the complex at 410 nm was continuously recorded.

3.1.2 Preparation of water samples and certified metal alloy solution

Sea and river water samples collected in Nalgene plastics were acidified by adding 1 mL of nitric acid (0.1 M) per 100 mL of sample solution after filtration over a 0.45 µm Millipore

Filter (Millford, MA). After the filtration and pre-treatment, water samples were injected directly into the FIA system for the determination of iron(III).

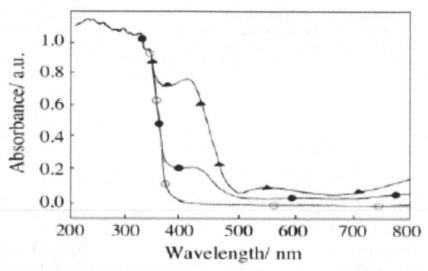

Fig. 5. Absorption spectra for DPA-4-SA and metal-DPA-4- SA complexes: (o) 5 × 10^{-5} M DPA-4-SA and 5× 10^{-5} M of each Co(II), Cu(II), Cr(III), Al(III), Cd(II), Ni(II), Mn(II), Ba(II), Ca(II), Ag(I), K(I), Na(I), Hg(II), Zn(II), and Mg(II); (•) 5 × 10^{-5} M of Fe(II)–DPA-4-SA; and (▲) 5 × 10^{-5} M of Fe(III)–DPA-4-SA.

Oxidizing iron(II) to iron(III) was used to determine the total iron amount. Hydrogen peroxide was chosen as the oxidizing agent for the determination of total iron. A concentration of 0.25 mol L^{-1} of H_2O_2 ensured total oxidation of iron(II) to iron(III) (Pons et al., 2005). Before the determination, H_2O_2 (10 mass %) was added to the water sample solution for complete oxidation of iron(II) to iron(III). Then, 20 µL of this solution were injected into the system, as in the procedure described above. Analyses were performed with the least possible delay.

A 0.10 g sample of the certified metal alloy (Zn/Al/Cu 43XZ3F) was dissolved in 12 mL of concentrated HCl + HNO3 (3 : 1) in a 100 mL beaker. The mixture was heated on a hot plate nearly to dryness; 5 mL of HNO3 were added to complete the dissolution and were diluted to 100 mL with deionized water. The solution was filtered and transferred quantitatively to a 1000 mL volumetric flask and filled up to the volume with deionized water. The volume of 10 mL of this solution was treated with H_2O_2 (10 mass %) for iron(II) oxidation. After the oxidation step, the solution was diluted 100 fold, and then, 20 µL of this solution were used for the determination of total iron.

3.2 Results and discussion

According to the spectrophotometric studies, iron (II) and iron(III) react with DPA-4-SA in aqueous medium to form complexes. As shown in Fig. 5, the absorption spectra corresponding to solutions of 5×10^{-5} M of each metal complex in water demonstrate strong absorption for the Fe(III)-DPA-4-SA complex. Iron(III) reacts with DPA-4-SA in the pH range of 2.0–6.0 forming a complex with absorption maxima at 410 nm and molar absorptivity of 1.60×10^4 L mol^{-1} cm^{-1}. The Fe(II)-DPA-4-SA complex presents only a slight absorption at this wavelength. Seventeen different metals do hardly react with DPA-4-SA in aqueous medium to from complexes. This can be an important advantage when developing a simplified FIA method for iron(III). Therefore, the specific absorbance maximum of the Fe(III)-DPA-4-SA complex at this wavelength can be applied in the selective determination of iron(III) in the flow-injection system.

The optimum experimental conditions were determined using a standard iron(III) solution. The concentration of DPA-4-SA, pH, and the flow-rate were the main variables influencing the intensity of the signal in the FIA system. Optimization of the FIA system was therefore performed by changing these variables one by one while applying 10 µg L^{-1} and 90 µg L^{-1} of iron(III) standard solutions in order to obtain the highest signal and better reproducibility at different concentration levels.

Influence of the DPA-4-SA concentration in the carrier solution on the peak height was examined by changing the DPA-4-SA concentration in the range of 1×10^{-2} M to 5×10^{-4} M in an acetate buffer solution (pH = 5.5), at the flow rate of 1 mL min^{-1}. Maximum peak heights were found using 1×10^{-3} M of the DPA-4-SA solution for both, 10 µg L^{-1} and 90 µg L^{-1}, iron(III) levels. Therefore, 1×10^{-3} M of DPA-4-SA was chosen as the color-developing component of the carrier solution.

The effect of flow-rate on the peak height of 10 µg L^{-1} and 90 µg L^{-1} iron(III) was examined by varying flow-rates from 0.2 mL min^{-1} to 2.0 mL min^{-1}. Peak heights decreased at flow-rates above 1.2 mL min^{-1} and below 0.7 mL min^{-1}. Flow-rates below 0.7 mL min^{-1} peaks also broadened. In the flow-rates range of 0.7–1.2 mL min^{-1}, there were slight differences in the peak heights. However, taking into consideration the stability of the pump, peak shape, and sampling time, the flow-rate of the reagent carrier solution was adjusted to 1 mL min^{-1}. This provided a sampling frequency of 60 h^{-1}.

pH of the carrier solution consisting of 1×10^{-3} M of DPA-4-SA was adjusted by adding simple acids and bases into the buffer to obtain the pH range of 3.30– 6.10. The peak shape and height were found maximum at pH 5.5. Therefore, 1×10^{-2} M of the acetate buffer solution at pH 5.5 was used throughout the study.

Reaction coil was used for the interaction of iron(III) and DPA-4-SA in the flow-injection system. The effect of the reaction coil (RC) length was examined by changing the coil length from 10 cm to 150 cm. The peak height decreased with the increase of length due to fast kinetics of the color forming reaction. The 10 cm length reaction coil was chosen since it produced the best peak height together with a good reproducibility.

The calibration graph for the determination of iron(III) was maintained under the optimized conditions as described above. A good linear relationship was observed for iron(III) ranging from 5 µg L^{-1} to 200 µg L^{-1}. The calibration curve equation was $A = 0.4018C + 2.0196$; $r2 = 0.9958$; $n = 6$, where A represents the absorbance measured as peak height and C the iron concentration in µg L^{-1}. The confidence limits of the intercept and the slope were calculated at the 95 % confidence level. The same calibration graph can be used for the determination of total iron. The detection limit estimated (S/N = 3) was 1 µg L-1 of iron(III).

The limit of quantification(LOQ) was calculated as recommended (Currie, 1995); based on a ten fold of the standard deviation of 10 consecutive injections of the blank, the value of 1.65 μg L^{-1} was obtained. The reproducibility of the method calculated as the relative standard deviation (RSD) of peak heights obtained from 5 injections of 10 μg L^{-1} iron(III) was 3.5 %.

Possible interferences in the determination of iron(III) were examined under the optimum experimental conditions. The effect of potential interfering ions on the determination of iron was investigated at the 5 % interference level. To carry out this study, 20 μL of a 20 μg L^{-1} iron(III) standard were injected. Table 4 summarizes the tolerance limits of the interfering ions. Most of the ions examined did not interfere with the iron(III) determination up to at least a 50000 fold excesses. The only interfering ion was iron(II), even 2 mg L^{-1} of iron(II) gave a positive interference.

Tolerance limit (mg L^{-1})	Foreign ion
Over 1000	Co(II), Cr(III), Al(III), Cu(II), Cd(II), Ni(II), Pb(II), Sn(II), Mn(II), Zn(II), K(I), Na(I), Ag(I), Ca(II), Mg(II), Ba(II), Hg(II), CN$^-$, NO$_3^-$, NO$_2^-$, SO$_4^{2-}$, CO$_3^{2-}$, Cl$^-$, Br$^-$, PO$_4^{3-}$, NH$_4^+$
Over 2	Fe(II)

Table 4. Effect of foreign ions on the determination of 20 μg L^{-1} of iron(III) in solution

The proposed method was applied in the determination of total iron in river and seawater samples. Iron(III) and total iron were determined according to the FIA procedure as described in the experimental section. The results obtained by both, standard addition and calibration curve, methods were in good agreement with each other. Atomic absorption measurements taken in water samples 1 and 2 are also given for comparison (Table 5).

Sample	Fe(III)[2] (μg L^{-1})		Total iron[2] (μg L^{-1})		Total iron[2] (μg L^{-1})
	Found[3]	Found[4]	Found[3]	Found[4]	AAS
Seaport (Sea water)	45.16 (0.06)	45.92 (0.21)	53.46 (0.19)	53.78 (0.27)	54.93(0.24)
Industry (Sea water)	56.28 (0.18)	56.11 (0.14)	76.45 (0.27)	76.13 (0.15)	78.19(0.16)
Atakum (River water)	21.45 (0.05)	21.18 (0.12)	32.69 (0.08)	31.85 (0.24)	34.47(0.36)
Mert (River water)	38.17 (0.11)	38.12 (0.19)	1.18 (0.04)	41.27 (0.16)	43.76(0.32)

1. Samples were collected at Samsun, Turkey.
2. Values in parantheses are the relative standard deviations for n =5 with confidence level of 95 %.
3. Calibration curve method.
4. Standard addition method.

Table 5. Analytical results of iron(III) and total iron in natural water samples[1]

The analytical value of total iron in water is in good agreement with that obtained by the AAS method. The accuracy of the proposed method was tested by the analysis of a certified metal alloy solution (MBH Zn/Al/Cu 43XZ3F). Three replicates of the solution using the sampling volume of 20 μL were analyzed. The certified and the obtained values were 0.085 % and (0.084 $\pm$ 0.006) of iron, respectively. An excellent agreement between the found and the certified values has been obtained for the certified metal alloy solution. The results obtained show that the proposed method can be applied in the determination of iron(III) and total iron content in water samples without a preconcentration process.

4. Flow injection spectrofluorimetric determination of iron (III) in water using salicylic acid (Reproduced with permission from the paper of Asan Adem et al., 2010. Copyright of Institute of Chemistry, Slovak Academy of Sciences)

In general terms, sensitivity of the spectrofluorimetric method is much higher than that of the spectrophotometric method. However, fluorescence reagents and methods suitable for the determination of iron are scarce and they suffer from serious interference of some metal cations such as aluminium, copper, and tin or they require a matrix separation step. Also, the reagents used for the determination of iron have a risk of toxicity (Tamm & Kalb, 1993; Yan et al., 1992; Cha et al., 1996; Ragos et al., 1998). Therefore, it is still important to develop simple and economical procedures that could be directly applied to real samples without the matrix separation step and with minimized reagent consumption.

In literature (Cha et al., 1998), salicylic acid has been used as a fluorescence reagent for the spectrofluorimetric determination of iron(III) in batch conditions. Experimentally it was found to be a very sensitive emission reagent for the spectrofluorimetric determination of iron(III) in the absence of iron (II). A very strong emission peak of salicylic acid in aqueous solution, which decreased linearly with the addition of iron(III), occurred at 409 nm with excitation at 299 nm. Also, salicylic acid is a commercially available reagent and it does not have a risk of serious toxicity when compared to the reagents used previously.

A simple and fast flow injection fluorescence quenching method for the determination of low levels of iron(III) in water has been developed. For this purpose, a preconcentration minicolumn consisting of cation-exchange resin was coupled to the FIA system. The use of mini-column in the system provided an improvement in sensitivity and the developed FIA method was successfully applied to the on-line determination of low levels of iron in real samples without the pre-concentration process. Fluorimetric determination was based on the measurement of the quenching effect of iron on salicylic acid fluorescence. An emission peak of salicylic acid in aqueous solution occurs at 409 nm with excitation at 299 nm. The effect of interferences from various metals and anions commonly present in water was also studied. The method was successfully applied to the determination of low levels of iron in real samples (river, sea, and spring waters).

4.1 Experimental

Analytical reagent grade chemicals were employed for the preparation of the standard, and the solutions were prepared using double distilled water. Standard iron(III) and iron (II) stock solutions (5×10^{-3} mol L^{-1} Fe(III) and Fe(II)) were prepared by dissolving $FeNH_4(SO_4)_2 \cdot 12H_2O$ and $Fe(NH_4)2(SO_4)_2 \cdot 6H_2O$ in water and were standardized by titration with EDTA. Iron(II) and iron(III) working standard solutions were prepared by appropriate dilution of the stock solutions with water immediately before use. Hydrogen peroxide solution, 30 mass %, was purchased from Merck (Darmstadt, Germany). Standard solutions of other metal ions (all of them from Merck (Darmstadt, Germany)) at different concentrations were prepared with doubly distilled water.

Buffer solution, 0.1 mol L^{-1} NH_4^+ /NH_3 at pH: 8.5, was used to produce analytical signal in the FIA system. Salicylic acid was provided from Merck (Darmstadt, Germany). Standard salicylic acid solutions were prepared daily by dissolving the appropriate amount of salicylic acid in an ethanol:water mixture (30 : 70). The reagent carrier solution was composed of 2×10^{-6} mol L^{-1} salicylic acid and 0.1 mol L^{-1} NH_4^+ /NH_3 buffer solution (90:10) at pH 8.5.

Fluorescence measurements for the batch experiments were performed with an SPF-500 model spectrofluorometer (American Instrument Co, Jessup, USA) using 1 cm quartz cells. Instrument excitation and emission slits were fixed at 10 nm. The light source was a 150 W Xenon lamp (American Instrument Co, Jessup, USA). Excitation and emission wavelengths were set at 299 nm and 409 nm, respectively. An eight-channel ISMATEC IPC peristaltic pump (Z̈urich, Switzerland), 0.75 mm i.d. PFTE tubing, was used to propel the samples and reagent solutions. Samples were injected into the carrier stream by a Rheodyne injection valve provided with a 20 µL loop. A Varian 2070 spectrofluorometer (Tokyo, Japan) using a 15 µL flow cell was used for the on-line measurements of analytical signals. Instrument excitation and emission slits were set at 20 nm. The light source was an ozoneless 75 W Xenon lamp (Tokyo, Japan). A strip chart recorder was attached to the instrument. Cation-exchange resin, sodium form of A650 W (100–200 mesh), was provided by the BioRad Labs (Hercules, CA, USA). The cation-exchange resin minicolumn (6 cm long, 2 mm i.d) was prepared in our laboratory.

pH measurements were carried out using a Jenway digital pH-meter model 3040 (Essex, England). An ATI UNICAM 929 model AAS (Cambridge, UK) flame atomic absorption spectrophotometer with a deuterium-lamp background correction was used for the determination of iron in reference to the FIA method. The measuring conditions were as follows: UNICAM hollow cathode lamp, 10 cm 1-slot burner, air–acetylene flame (fuel gas flow-rate of 1.50 L min^{-1}), 0.2 nm spectral bandwidth, and 7 mm burner height. The wavelength and the lamp current of iron were 248 nm and 5 mA, respectively. The flow injection manifold was similar to that proposed in our previous study (Isildak et al., 1999). Peristaltic pump was used to transport the reagent carrier solution through the system. The sample was injected using an injection loop (20 µL). The reagent carrier solution and the sample were allowed to mix in the flow stream and in the mini-column. The decrease in the fluorescence intensity of the salicylic acid as a function of Fe(III) concentration was measured in the flow cell using 299 nm for excitation and 409 nm for emission. Water samples were obtained from different places of the river, sea and thermal spring in Samsun, Turkey. They were filtered through a 0.45 µm Millipore Filter (Millford, MA, USA). Water samples were split into two portions: one part was directly injected into the FIA system for the determination of iron(III). Before the analysis of the other part, 1 mL of H_2O_2 (10 mass %) was added to a 9 mL sample solution for complete oxidation of iron(II) to iron(III). Then, 20µL of this solution were injected into the system for the determination of total iron, as in the procedure described above.

A 0.10 g sample of the certified metal alloy (Zn/Al/Cu 43XZ3F) was dissolved in 12 mL of concentrated HCl + HNO_3 (3 : 1) in a 100 mL beaker. The mixture was heated on a hot plate nearly to dryness; 5 mL of HNO_3 were added to complete the dissolution, and the solution was diluted to 100 mL with deionized water, filtered and transferred quantitatively to a 1000 mL volumetric flask and filled up to the volume with deionized water. The volume of 10 mL of this solution was treated with H2O2 (10 mass %) for iron(II) oxidation. After the oxidation step, the solution was diluted 100 fold, and then, 20 µL of this solution were used for the determination of total iron.

4.2 Results and discussion

Fig. 6 shows the fluorescence emission spectra of $5×10^{-5}$ mol L^{-1} salicylic acid in a buffer solution at pH 8.5 before and after the reaction with $1×10^{-5}$ mol L^{-1} iron(II) and iron(III), respectively, in batch experiments. As can be seen, the intensity of salicylic acid fluorescence decreased significantly in the presence of iron(III). From these spectra, the emission wavelength chosen for the FIA measurement was 409 nm, using 299 nm for the fluorescence excitation.

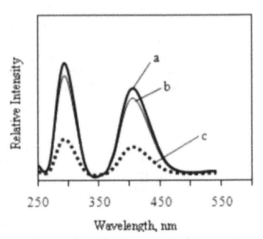

Fig. 6. Emission spectrum of $5×10^{-5}$ M salicylic acid in batch experiment (in the absence and presence of $1×10^{-5}$ M Fe(III) and $1×10^{-5}$ M Fe(II) ions): a) salicylic acid, b) salicylic acid + Fe(II), c) salicylic acid + Fe(III).

4.2.1 Optimization of FI manifold

Optimization of the flow system was performed to establish the best FIA variables. A fixed standard Fe (III) solution, 10 µg L^{-1} was injected into the flow system for the determination of optimum experimental conditions. The main variables influencing the intensity of the signal were: flow-rate, pH, and the concentration of salicylic acid. Therefore, optimization of the FIA system was carried out by changing these variables one by one.

The effect of salicylic acid in the carrier solution on the peak height was examined by changing the amount of salicylic acid in the range of $5×10^{-7}$–$5×10^{-5}$ mol L^{-1} in buffer solution at pH 8.5, at the flow rate of 1.0 mL min^{-1}. Peak heights were found maximum using a $2×10^{-6}$ mol L^{-1} salicylic acid solution for 10 µg L^{-1} iron(III) levels. Therefore, $2×10^{-6}$ mol L^{-1} salicylic acid was chosen as the fluorescence reagent in the carrier solution.

The effect of flow-rate on the peak height of iron(III) was examined by varying the flow-rate from 0.5 mL min^{-1} to 1.5 mL min^{-1}. Peak heights decreased at flow-rates above 1.2 mL min^{-1}

and below 0.8 mL min^{-1}. Below 0.8 mL min^{-1} the peaks also broadened. Between the flow-rates of 0.8–1.2 mL min^{-1}, there were slight differences in the peak heights. Considering the stability of the pump, peak height, and sampling time, the flow-rate of the reagent carrier solution was adjusted to 1.0 mL min^{-1}. This provided the sampling frequency of 60 h^{-1}. pH of the carrier solution consisting of 2×10^{-6} mol L^{-1} salicylic acid was adjusted by an NH$_4^+$ /NH$_3$ buffer solution to obtain the pH range of 8.0–10.0. The peak heights were found maximum at pH 8.5. Therefore, a 0.1 mol L^{-1} NH$_4^+$ /NH$_3$ buffer solution (90 : 10) at pH 8.5 was used throughout the study.

The use of a mini-column in the flow-injection system provided an improvement in the sensitivity and selectivity due to on-line pre-concentration and fast interaction of metal ions with reagent molecules in the carrier solution (Isildak et al., 1999). A mini-column packed with strong cation-exchange resin was selected because metal ions are strongly bound by the resin so that low amounts of the resin can be used. Higher amounts of the resin minimized the use of higher flowrates due to an increase in the hydrodynamic pressure. Sampling time in the FIA system depends on the retention time in the cation exchange mini-column and the residence time in the tubing in the flow-path. The effect of the column length was examined by changing the column length between 2 cm and 10 cm. From the results obtained, 6 cm column length brought the best results for the peak shape and sensitivity for iron for all concentration levels studied.

Also a mixing coil and a mini-column packet with silica and glass beads were inserted into the analytical path instead of the cation-exchange resin minicolumn. However, the observed peak height and sensitivity for iron(III) were lower and poorer, for all concentration levels studied. This result can originate from the short remaining time of iron(III) in each column, which means a narrow interacting zone of the sample. Finally, a mini-column packed with strong cation-exchange resin was used throughout the study for the determination of iron(III). Indeed, a significant improvement of the selectivity and sensitivity was observed.

4.2.2 Analytical performance characteristics

Analytical performance characteristics of the method were evaluated under optimum conditions. Fig. 7 shows typical flow signals for iron(III) obtained by the proposed method. The reaction of iron(III) with salicylic acid resulted in negative peaks due to the fluorescence quenching of salicylic acid. Under the optimum working conditions, calibration graphs were prepared from the results of triplicate measurements of iron(III) standard solutions of increasing concentration. The calibration graph showed a good linearity from 5–100 µg L^{-1} iron(III) with the linear regression equation: Y = 0.0353X + 0.0909, where Y is the peak height (cm) and X is the concentration of iron(III) in µg L^{-1}. The correlation coefficient was r^2 = 0.9963 and the relative Standard deviation (RSD) of the method based on five replicate measurements of 10 µg L^{-1} iron(III) was 1.25 % for a 20 µL injection volume. The limit of detection (determined as three times the standard deviation of the blank) was 0.3 µg L^{-1} and the sampling rate was 60 h^{-1}. The limit of quantification (LOQ) was calculated as recommended (Currie, 1995); based on a ten fold standard deviation of ten consecutive injections of the blank, the value of 1.12 µg L^{-1} was obtained.

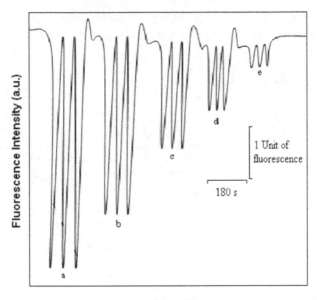

Time (s)

Fig. 7. Flow signal for iron(III) standard solutions by fluorescence quenching-FIA a) 100 µg L−1, b) 75 µg L−1, c) 50 µg L−1, d) 25 µg L−1, and e) 5 µg L−1 when using the optimized FIA system.

4.2.3 Interference study

The effect of diverse ions on the detection of iron by the present system were examined using a solution containing 10 µg L^{-1} iron(III) and one of the other ions. The tolerable concentration of each diverse ion was taken as the highest concentration causing the error of ± 5 %. The results are summarized in Table 6.

Tolerance limit (mg L^{-1})	Foreign ion
No interfere	CO_3^{2-}, SCN^-, Br^-, SO_4^{2-}, Ca^{2+}, Zn^{2+}
Over 50 000	Co(II), Cr(III), Al(III), Cu(II), Cd(II), Ni(II), Pb(II), Mn(II), K(I), Na(I), Ag(I), Mg(II), Ba(II), Hg(II), CN^-, NO_3^-, NO_2^-, Cl^-, PO_4^{3-}, NH_4^+
Over 100	Fe(II)

Table 6. Effect of foreign ions on the determination of 10 µg L^{-1} of iron(III) in solution

4.2.4 Analysis of water samples

The proposed method was applied to the determination of iron in river, sea, and thermal spring water samples to evaluate its applicability. Iron(III) and total iron were determined according to the FIA procedure as described in the experimental section. Table 7 shows the analytical results of iron(III) and total iron. Atomic absorption measurements taken were also given for comparison. The results obtained with the standard addition and the calibration curve methods, and the AAS measurements were in good agreement with each other.

Sample	Fe(III)[2] (µg L[-1])		Total iron [2] (µg L[-1])			
	Found[3]	Found[4]	Found[3]	Found[4]	AAS	Ec (%)
Seaport (Sea water)	52.16 (0.12)	52.92 (0.21)	67.25 (0.10)	67.52 (0.12)	66.98 (0.05)	0.60
Atakum(River water)	25.41 (0.10)	26.01 (0.19)	37.41 (0.14)	38.01 (0.17)	38.15 (0.07)	1.16
Kurtun river	32.84 (0.24)	33.57 (0.28)	48.14 (0.19)	48.57 (0.27)	49.12 (0.09)	1.55
Spring water (1)	10.95 (0.15)	11.25 (0.27)	16.75 (0.32)	16.20 (0.28)	16.62 (0.18)	0.88
Spring water (2)	12.65 (0.09)	13.18 (0.12)	21.83 (0.08)	21.32 (0.24)	21.75 (0.14)	0.81
Spring water (3)	38.17 (0.11)	38.12 (0.19)	52.54(0.04)	52.73 (0.16)	52.95 (0.12)	0.60

1. Samples were collected at Samsun, Turkey.
2. Values in parantheses are the relative standard deviations for n =5 with confidence level of 95 %.
3. Calibration curve method.
4. Standard addition method.

Table 7. Determination of total iron in water samples[1]

Accuracy of the proposed method was also tested by analyzing a certified metal alloy solution (MBH Zn/Al/Cu 43XZ3F). Three replicates of the solution using the sampling volume of 20 µL were analyzed. The certified and the obtained values were 0.085 % and (0.084 ± 0.006) % of iron, respectively. An excellent agreement between the found and the certified values was obtained for the certified metal alloy solution. The obtained results show that the proposed method can be applied to the determination of iron(III) and total iron content in water samples without a pre-concentration process.

5. A simple flow injection spectrophotometric determination method for iron (III) based on O-acetylsalicylhydroxamic acid complexation (Reproduced with permission from the paper of Andac Muberra et al., 2009. Copyright of Institute of Chemistry, Slovak Academy of Sciences)

1,10-phenanthroline and salicylic acid are the most reported chelating agents applied for the determination of iron(III) and total iron after oxidation to iron(III) (Tesfaldet et al., 2004; Udnan et al., 2004). A number of other chelating agents that have been reported for the spectrophotometric and/or flow-injection spectrophotometric determination of iron(III) and total iron include 2-thiobarbituric acid (Morelli, 1983), norfloxacin (Pojanagaron et al., 2002) tiron (van Staden & Kluever, 2002) DMF (Asan et al., 2003), tetracycline (Sultan et al., 1992) and chlortetracycline (Wirat, 2008). Flow-injection spectrophotometric methods based on the above chelating agents are either not selective, or a masking agent has to be used. However, highly selective, simple and economical methods for routine determination of iron(III) in different sample matrices are still required. In the present study, a simple and rapid flow-injection spectrophotometric method for the determination of iron (III) and total iron is proposed. The method is based on the reaction between iron (III) and O-acetylsalicylhydroxamic acid (AcSHA) in a 2 % methanol solution resulting in an intense violet complex with strong absorption at 475 nm. The reagent itself is sparingly soluble in water and did not absorb in the visible region of the spectrum, therefore, it might be well suited for flow-injection analysis of iron(III) and total iron. An addition of copper sulphate (1×10^{-4} mol L^{-1}) into the reagent carrier solution resulted in baseline absorbance, and possible interfering ions were eliminated without a significant decrease in the sensitivity of the method. The method was successfully applied in the determination of iron (III) and total iron in water and ore samples. The method was verified by analysing a certified reference material Zn/Al/Cu 43XZ3F and also by the AAS method.

5.1 Experimental

All chemicals used were of analytical reagent grade, and solutions were prepared from double deionised water. Standard iron(II) and iron(III) stock solutions were prepared by dissolving 278.02 mg of iron(II) and 489.96 mg of iron(III) sulphate (Merck; Darmstadt, Germany) in 100 mL of 0.01 mol L^{-1} hydrochloric acid to give 0.01 mol L^{-1} stock solution of iron(II) and iron(III). Iron(II) and iron(III) working standard solutions were prepared daily by suitable dilution of the stock solutions with double deionised water. Standard reference material consisting of 0.085 % Fe (Zn/Al/Cu 43XZ3F) was provided from MBH Analytical Ltd. (UK). Hydrogen peroxide solution of 30 vol. % was obtained from Merck. AcSHA was synthesised according to the procedure described previously (Asan et al., 2003). A stock solution of AcSHA (0.01 mol L^{-1}) was prepared by dissolving 0.095 g of AcSHA in 100 mL of aqueous methanol (2 vol. %). For the spectrophotometric study, AcSHA complex solutions of various metals were prepared by mixing 1 mL of 1×10^{-4} mol L^{-1}standard solution of each metal in double deionised water with the suitable volume of 1×10^{-4} mol L^{-1} AcSHA stock solution. Reagent carrier solution was composed of AcSHA in a 2 % methanol solution and 1×10^{-4} mol L^{-1} $CuSO_4$ in 0.001 mol L^{-1} HCl 98 % (pH 2.85). UV-VIS spectra of metal-AcSHA complexes were taken with a Unicam spectrophotometer (GBC Cintra 20, Australia). A Jenway 3040 Model digital pH-meter was used for the pH measurements. In the FIA system, a peristaltic pump (ISMATEC; IPC, Switzerland) 0.50 mm i.d. PTFE tubing was used to propel the samples and reagent solutions. Samples were injected into the carrier stream by a 7125 model stainless steel high pressure Rheodyne injection valve provided with a 20 µL loop. Absorbance of the coloured complex formed was measured with a UV-VIS spectrophotometer equipped with a flowthrough micro cell (Spectra SYSTEM UV 3000 HR,Thermo Separation Products, USA), and connected to a computer incorporated with a PC1000 software programme. A UNICAM 929 model (Shimadzu AA-68006) flame atomic absorption spectrophotometer with a deuterium-lamp background correction was used for the determination of iron in reference to the FIA method. The measuring conditions were as follows: UNICAM hollow cathode lamp, 10 cm 1-slot burner, air-acetylene flame (fuel gas flow-rate 1.50 L min^{-1}), 0.2 nm spectral bandwidth, and 7 mm burner height. The wavelength and the lamp current of iron were 248 nm and 5 mA, respectively.

The FIA system used, similar to that proposed in our previous works (Asan et al., 2003), is quite simple. The sample solution was introduced into the reagent carrier solution by the Rhodyne injection valve. A water-soluble complex (λmax = 475 nm) was then formed on the passage of the reagent carrier solution in the mixing coil. As a mixing coil, PTFE tubing (50 cm long) was attached before the flow-through detection cell. The absorbance of the coloured complex was selectively monitored in the cell at 475 nm. The transient signal was recorded as a peak, the height of which was proportional to the iron(III) concentration in the sample, and it was used in all measurements. Five replicate injections per sample were made.

Sea and river water samples collected in Nalgene plastics were acidified by adding 1 mL of nitric acid (0.1 mol L^{-1}) per 100 mL of sample solution after filtration over a 0.45 μm Millipore Filter (Millford, MA). After the filtration, water samples were injected directly into the FIA system for the determination of iron(III).

Total iron was determined by oxidising iron(II) to iron(III). Hydrogen peroxide was chosen as the oxidising agent for the determination of total iron. A 0.25 mol L^{-1} H$_2$O$_2$ concentration ensured total oxidation of iron(II) into iron(III) (Pons, et al., 2005). Before the determination of total iron, H$_2$O$_2$ (10 mass %) was added to the water sample solution for complete oxidation of iron(II) to iron(III). Then, 20 μL of this solution were injected into the system, as in the procedure described above. A 0.10 g sample of the certified metal alloy (Zn/Al/Cu 43XZ3F) was dissolved in 12 mL of concentrated HCl and HNO$_3$ (3 : 1) in a 100 mL beaker. The mixture was heated on a hot plate nearly to dryness; 5 mL of HNO$_3$ were added to complete the dissolution and the solution was diluted to 100 mL with deionised water. The solution was filtered and transferred quantitatively to a 1000 mL volumetric flask and filled up to volume with deionised water. 9 mL of this solution were treated with 1 mL of H$_2$O$_2$ (10 mass %) for iron(II) oxidation. After the oxidation step, 20 μL of this solution were used in the determination of total iron. Metal ore samples (0.10 g) were powdered (≥ 500 mesh) and prepared as in the procedure described above. All analyses were performed with the least possible delay.

5.2 Results and discussion

5.2.1 Spectrophotometric studies of AcSHA-metal complexes

Metal ions react with AcSHA in aqueous media in the range of pH 2.0–10.0 forming coloured complexes with different stoichiometry. These complexes are fairly soluble in aqueous media (O'Brien et al., 1997). Their absorption spectra corresponding to solutions of 5 × 10^{-5} mol L^{-1} metal complexes measured against a reagent blank are shown in Fig. 8.

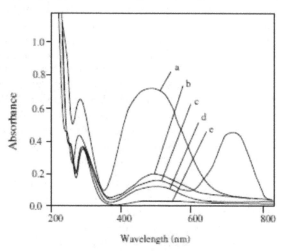

Fig. 8. Absorption spectras of 5x10^{-5} M AcSHA and M-(AcSHA)$_n$ complexes. a) Fe(III)-(AcSHA)$_n$; b) Fe(II)-(AcSHA)$_n$; c) Cu-(AcSHA)$_n$; d) M-(AcSHA)$_n$; (M: Ni, Co, Zn, Pb); e) AcSHA only.

As can be seen from Fig. 8, only AcSHA reacted efficiently with iron to form iron-(AcSHA)n complexes with the absorbance maxima at 475 nm. At this wavelength, AcSHA itself has no absorption while Ac- SHA complexes of copper(II), nickel(II), cobalt(II), and zinc(II), among all metal ions with the anions tested, show a negligible absorption. The FIA setup shown in Fig. 9. was used in order to develop an FIA method based on the above phenomenon.

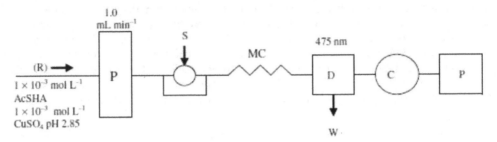

Fig. 9. Flow diagram of the flow-injection analysis system used for the determination of iron (III) and total iron, R; reagent carrier solution (1x10^{-4} M AcSHA, 1x10^{-4} M CuSO$_4$, pH: 2.85), P, Peristaltic pump, S; Rheodyne sample injection valve, MC; mixing coil (50 cm long, 0.5 mm i.d), D; spectrophotometric detector (λ_{max} = 475 nm), W; waste, C; computer, P; printer.

5.2.2 Optimisation of chemical variables and FIA manifold

Various variables closely related to iron determination were examined using a simple flow-injection analysis system with a fixed iron(III) concentration of 5 µg L^{-1}. The AcSHA concentration was varied from 1×10^{-5} mol L^{-1} to 1×10^{-2} mol L^{-1}. The peak height was found to increase with the AcSHA concentration increasing up to 1×10^{-4} mol L^{-1}, no noticeable increase was found at higher concentrations. Therefore, 1×10^{-4} mol L^{-1} AcSHA was used as the colour developing component of the carrier solution. With the concentration of AcSHA fixed at 1×10^{-4} mol L^{-1}, pH of the carrier solution was varied from 1.5 to 5.5. The interference effect of iron(II) was found to increase with pH increasing up to 3.5 and to remain constant at higher pH. Also, the peak heights were found to increase with pH increasing up to 3.0, to remain constant up to 4.0, and to decrease slightly above this value. pH of the reagent carrier was, however, adjusted to 2.85 to obtain the maximum peak height and minimum iron(II) interference in the analysis. To obtain a reasonable background of absorption and a smooth baseline, CuSO$_4$ was added into the carrier solution. The CuSO$_4$ concentration was varied from 1×10^{-5} mol L^{-1} to 1×10^{-2} mol L^{-1}. When the concentration of CuSO$_4$ was 1×10^{-4} mol L^{-1}, the baseline was stable and the interference effects of nickel(II), cobalt(II), and zinc(II) were found minimum. Over the CuSO$_4$ concentration of 1×10^{-4} mol L^{-1}, the sensitivity of the method decreased.

In order to proceed with the final system design, the effects of sample volume, mixing coil length and flow-rate were studied at the optimal pH (2.85), and fixed concentrations of AcSHA (1×10^{-4} mol L^{-1}) and CuSO$_4$ (1×10^{-4} mol L^{-1}). The sample volume was varied from 5–50 µL. The peak height was decreased by decreasing the sample size, and the peaks were broadened with the increasing sample size due to the sample zone dispersion. The sample injection volume of 20 µL was selected as a compromise between the sensitivity and sample throughput rate. The mixing coil (MC) was examined using PTFE tubing (0.5 mm i.d.) of different lengths ranging between 10 cm and 150 cm. The peak height increased with the increasing mixing coil length from 10–50 cm, decreased at lower concentrations and broadened at higher concentrations and longer coil lengths. The mixing coil length of 50 cm was chosen since it resulted in the best peak height and good reproducibility.

The flow-rate was varied from 0.2 mL min^{-1} to 2 mL min^{-1}. The peak height decreased with the increasing flow-rate, probably due to the extent of the reaction decrease. The flow-rate of 0.8 mL min^{-1} was selected as a compromise between the sample throughput rate and sensitivity. A linear calibration graph for 4–150 µg L^{-1} iron(III), with the regression coefficient of 0.9914, was obtained under optimum conditions. The relative standard deviation for the determination of 5 µg L^{-1} iron(III) was 0.85 % (10 replicate injections), RSD of the data was below 3 %. The limit of detection (blank signal plus three times the standard deviation of the blank) was 0.5 µg L^{-1}. The sample throughput of the proposed method was almost 60 h^{-1}.

Tolerance limit (µg L^{-1})	Foreign ion
Over 50000	Cr(III), Al(III), Cd(II), Mn(II), K(I), Na(I), Ag(I), Ca(II), Mg(II), Ba(II), Hg(II), CN$^-$, NO$_3^-$, NO$_2^-$, SO$_4^{2-}$, CO$_3^{2-}$, Cl$^-$, Br$^-$, PO$_4^{3-}$, NH$_4^+$
Over 100	Fe (II)

Table 8. Effect of foreign ions on the determination of 5 µg L^{-1} of iron (III) in solution

The interference effects of many cations and anions on the determination of 5 µg L^{-1} iron(III) were examined. The results summarised in Table 8 represent tolerable concentrations of each diverse ion taken as the highest concentration causing an error of 3 %. Most of the ions examined did not interfere with the determination of iron(III). The major interference was caused by iron(II) at the amount of 100 µg L^{-1}. It is known that zinc and cobalt are the main interference metal ions in the determination of iron (Ensafi et al., 2004). In this study, the interference of these ions was completely eliminated by an addition of copper sulphate ($1{\times}10^{-4}$ mol L^{-1}) to the reagent carrier solution. Background absorbance of copper(II) maintained in the reagent carrier solution eliminated possible interfering ions and improved the determination of iron(III). It is apparent from Table 1 that the proposed method tolerates all interfering species tested in satisfactory amounts, and it is therefore adequately selective for the determination of iron(III) and total iron.

5.2.3 Applications

The FIA method was applied in the determination of iron(III) and total iron in water and ore samples. In order to evaluate the accuracy of the proposed method, the determination of total iron in a standard reference material (Zn/Al/Cu 43XZ3F) and in a metal alloy sample was carried out. The analytical results obtained by the proposed method are in good agreement with the certified values as shown in Table 9.

For the application of the proposed FIA method to water samples; river and sea water samples collected from different sources were analysed using both the calibration curve and the standard addition methods. The values obtained from the calibration curve and the standard addition methods are in good agreement as shown in Table 10. Atomic absorption measurements taken in water samples are also given for comparison (Table 10). The analytical value of total iron in water is in good agreement with that obtained by the AAS method.

Sample	Total Fe[1] (%)	Certified Fe (%)
Alloy (1)	8.23(0.24)	8.58
Alloy (2)	16.15(0.17)	16.62
Std Zn/Al/Cu 43XZ3 F	0.083(0.022)	0.085

[1] Values in parenthesis are the relative standard deviations for n=5 with confidence level of 95 %.

Table 9. Total iron content of iron alloys and standard reference material

Samples(1)	Iron (III)(2) (μg L-1)		Total iron(2) (μg L-1)		
	Found(3)	Found(4)	Found(3)	Found(4)	AAS
Kurtun river	38.33(0.24)	38.55(0.12)	42.33(0.02)	42.91(0.18)	43.65(0.17)
Seaport	78.84(0.32)	78.65(0.24)	95.13(0.12)	95.75(0.06)	97.12(0.12)
Baruthane sea water	47.51(0.18)	47.62(0.14)	57.24(0.04)	57.65(0.15)	58.97(0.24)

(1) Samples were collected at Samsun, Turkey.
(2) Values in parenthesis are the relative standard deviations for n=5 with confidence level of 95 %.
(3) Calibration curve method.
(4) Standard addition method.

Table 10. Determination of iron (III) and total iron in river and sea water samples

The results obtained show that the proposed method can be applied in the determination of iron(III) and total iron content in water samples without a preconcentration process.

6. Conclusions

A number of highly sensitive, selective and rapid flow-injection spectrophotometric and spectrofluorimetric analysis methods for the determination of iron (II), iron (III) and total iron in a wide concentration range, without employing any further treatment, have been described. The methods were based on the reactions of iron (II) and iron (III) with different complexing agents in different carrier solutions in FIA. In addition to the simplicity and low reagent consumption of the methods, the complexing agents used are commercially available and may not have a risk of serious toxicity, thus enhancing the potential applicability of the methods for iron analysis in real samples. Several parameters affecting to the determination of iron (II) and iron (III) were examined. The methods developed have been successfully applied to the determination of iron (II), iron (III) and total iron different types of water samples including river, sea, industry and spring water samples. The methods were also verified by applying certified reference materials.

7. References

Ahmed M. J. and Roy U. K. (2009) A simple spectrophotometric method for the determination of iron(II) aqueous solutions. Turkish journal of chemistry. 33, 709-726

Alonso J., Bartroli J., Valle M. D. and Barber R. (1989) Sandwich techniques in flow-injection analysis Part 2. Simultaneous determination of iron(II) and total iron. Analytica Chimica Acta, 219, 345-350.

Ampan P., Lapanantnoppakhum S., Sooksamiti P., Jakmunee J., Hartwell S. K., Jayasvati S., Christian G. D. and Grudpan K. (2002) Determination of trace iron in beer using flow-injection systems with in-valve column and bead injection. Talanta, 58, 1327-1334.

Andac M., Asan A. and Isildak I. (2001) Flow-injection spectrophotometric determination of cobalt/II) at low µg l⁻¹ levels with 4-benzylpiperidinedithiocarbamate. Analytica Chimica Acta, 434, 143-147.

Andac M., Asan A. and Isildak I. (2009) A simple flow injection spectrophotometric determination method for iron(III) based on O-acetylsalicylhydroxamic acid complexation. Chemical Papers, 63, 268-273

Araujo A. N., Gracia J., Lima J. L. F. C., Poch M., Luica M. and Saraiva M. F. S. (1997) Colorimetric determination of iron in infant fortified formulas by sequential injection analysis. Fresenius' Journal of Analytical Chemistry, 357, 1153-1156.

Asan A., Andac M. and Isildak I. (2010) Flow injection spectrofluorimetric determination of iron(III) in water using salicylic acid. Chemical Papers, 64(4) 424-428.

Asan A., Andac M., Isildak I. and Tinkilic N. (2008) Flow Injection Spectrophotometric Determination of Iron(III) Using Diphenylamine-4-sulfonic Acid Sodium Salt. Chemical Papers, 62 (4), 345-349.

Asan A., Andac M. and Isildak I. (2003) Flow-injection spectrophotometric determination of nanogram levels of iron(III) with N,N-dimethylformamide. Anal. Sci., 19 (7), 1033-1036.

Asan A., Isildak I., Andac M. and Yilmaz F. (2003) A simple and selective flow-injection spectrophotometric determination of copper(II) by using acetylsalicylhydroxamic acid. Talanta, 60, 861-866.

Bagheri H., Gholami A. and Najafi A. (2000) Simultaneous preconcentration and speciation of iron(II) and iron(III) in water samples by 2-mercaptobenzimidazole-silica gel sorbent and flow injection analysis system. Analytica Chimica Acta, 424, 233-242.

Bruno H. A., Andrade F. J., Luna P. C. and Tudino M. B. (2002) Kinetic control of reagent dissolution for the flow injection determination of iron at trace levels. Analyst, 127, 990-994.

Busev A. I., Tiptsova V. G. and Ivanov V. M. (1981) Anal. Chem. of rare elements, 385.

Cha K. W. and Park C. I. (1996) Spectrofluorimetric determination of iron(III) with 2-pyridinecarbaldhyde-5-nitropyridylhydrazone   in   the   presence   of hexadecyltrimethylammonium bromide surfactant. Talanta, 43, 1335-1341.

Cha K.. W. and Park K.. W. (1998) Determination of iron(III) with salicylic acid by the fluorescence quenching method. Talanta, 46, 1567-1571.

Chen J. Q., Gao W. and Song J. F. (2006) Flow-injection determination of iron(III) in soil by biamperometry using two independent redox couples. Sensors and Actuators B, 113,194-200.

Bowie A. R., Achterberg E. P., Mantoura R. F. C. and Worsfold, P. J. (1998) Determination of sub-nanomolar levels of iron in seawater using flow injection with chemiluminescence detection. Analytica Chimica Acta, 36, 189–200.

Currie L. A. (1995) Nomenclature in evaluation of analytical methods including detection and quantiication capabilities. Pure and Applied Chemistry, 67, 1699–1723.

Dojlido J. R. and Best G. A. (1993) Chemistry of water and water pollution. Chichester: Ellis Horwood series in water and wastewater technology. Prentice Hall Inc. Englewood Cliffs, 21: 251.

Ensafi A. A., Chamjangali M. A. and Mansour H. R. (2004) Sequential determination of iron(II) and iron(III) in pharmaceutical by flow-injection analysis with spectrophotometric detection. Analytical Sciences, 20, 645–650.

Giokas D. L., Paleologos E. K.. and Karayannis M. I. (2002) Speciation of Fe(II) and Fe(III) by the modified ferrozine method, FIA-spectrophotometry, and flame AAS after cloudpoint extraction. Analytical and Bioanalytical Chemistry, 373, 237–243.

Guo T. and Baasner J. (1993) Determination of mercury in urine by flow-injection cold vapour atomic absorption spectrometry. Analytica Chimica Acta, 278, 189–196.

Hirata S., Yoshihara H. and Aihara M. (1999) Determination of iron(II) and total iron in environmental water samples by flow injection analysis with column preconcentration of chelating resin functionalized with Nhydroxyethylethylenediamine ligands and chemiluminescence detection. Talanta, 49, 1059–1067.

Isildak I., Asan A. and Andac M. (1999) Spectrophotometric determination of copper(II) at low µg l−1 levels using cationexchange microcolumn in flow-injection. Talanta, 48, 219– 224.

Kawakubo S., Natio A., Fujihara A. and Iwatsuki M. (2004) Field determination of trace iron in fresh water samples by visual and spectrophotometric methods. Analytical Sciences, 20, 1159–1163.

Kass M. and Ivaska A. (2002) Spectrophotometric determination of iron(III) and total iron by sequential injection analysis technique. Talanta, 58, 1131–1137.

Lunvongsa S., Takayanagi T., Oshima M. and Motomizu S. (2006) Novel catalytic oxidative coupling reaction of N,N-dimethyl-p-phenylenediamine with 1,3-phenylenediamine and its applications to the determination of copper and iron at trace levels by flow injection technique. Analytica Chimica Acta, 576, 261–269.

Lunvongsa S., Oshima M. and Motomizu S. (2006) Determination of total and dissolved amount of iron in water samples using catalytic spectrophotometric flow injection analysis. Talanta, 68, 969–973.

MacCarthy P. and Zachary D. H. (1986) Journal of Chemical Education 63 (3): 162–167.

Molina-Diaz A., Ortega-Carmona I. and Pascual-Reguera M.I. (1998) Indirect spectrophotometric determination of ascorbic acid with ferrozine by flow-injection analysis. Talanta, 47:531-536

Mohammad R. S., Deepika J., Jonathan P. K. and Keith A. H. (2004) Determination of labile Cu^{2+} in fresh waters by chemiluminescence: interference by iron and other cations. Talanta, 62: 924-930

Morelli B. (1983) Determination of iron(III) and copper(II) by zeroth, first and second derivative spectrophotometry with 2-thiobarbituric acid (4,6-dihydroxy-2-mercaptopyrimidine) as reagent. Analyst, 108, 870–879.

Mulaudzi L. V., van Standen J. F. and Stefan R. J. (2002) On-line determination of iron(II) and iron(III) using a spectrophotometric sequential injection system. Analytica Chimica Acta, 467, 35–49.

Muller H., Muller V. and Hansen E. H. (1990) Simultaneous differential rate determination of iron(II) and iron(III) by flow-injection analysis. Analytica Chimica Acta, 230, 113–123.

O'Brien E. C., Roy S. L., Levaillain J., Fitzgerald D. J. and Nolan K. B. (1997) Metal complexes of salicylhydroxamic acid and O-acetylsalicylhydroxamic acid. Inorganica Chimica Acta, 266, 117–120.

Ohno S., Tanaka M., Teshima N. and Sakai T. (2004) Successive determination of copper and iron by a flow injectioncatalytic photometric method using a serial flow cell. Analytical Sciences, 20, 171–175.

Qin W., Zhang Z. J. and Wang F. C. (1998) Chemiluminescence flow system for the determination of Fe(II) and Fe(III) in water. Fresenius' Journal of Analytical Chemistry, 360, 130–132.

Pascual-Reguera M. I., Ortega-Carmona I. and Molina-Diaz A. (1997) Spectrophotometric determination of iron with ferrozine by flow-injection analysis. Talanta, 44, 1793–1801.

Pojanagaron T., Watanesk S., Rattanaphani V. and Liawruangrath S. (2002) Reverse flow-injection spectrophotometric determination of iron(III) using norfloxacin. Talanta, 58, 1293–1300

Pons C., Forteza R. and Cerd`a V. (2005) Multi-pumping flow-system for the determination, solid-phase extraction and speciation analysis of iron. Analytica Chimica Acta, 550, 33–39.

Pons C., Forteza R. and Cerd`a V. (2005) The use of anion exchange disks in an optrode coupled to a multi-syringe flowinjection system for the determination and speciation analysis of iron in natural water samples. Talanta, 66, 210–217.

Pulido-Tofino P., Barrero-Moreno J. M. and Perez-Conde M. C. (2000) A flow-through fluorescent sensor to determine Fe(III) and total inorganic iron. Talanta, 51, 537–545.

Ragos G. C., Demertzis M. A. and Issopoulos P. B. (1998) A high-sensitive spectrofluorimetric method for the determination of micromolar concentrations of iron(III) in bovine liver with 4-hydroxyquinoline. II Farmaco, 53, 611–616.

Reguera M. I. P., Carmona I. O. and Diaz A. M. (1997) Spectrophotometric determination of iron with ferrozine by flow-injection analysis. Talanta, 44, 1793–1801.

Saitoh K.., Hasebe T., Teshima N., Kurihara M. and Kawashima T. (1998) Simultaneous flow-injection determination of iron(II) and total iron by micelle enhanced luminol chemiluminescence. Analytica Chimica Acta, 376, 247–254.

Sangi M. R., Jayatissa D., Kim J. P. and Hunter K. A. (2004) Determination of labile Cu2+ in fresh waters by chemiluminescence: interference by iron and other cations. Talanta, 62, 924–930.

Sultan S. M., Suliman F. O., Duffuaa S. O. and Abu-Abdoun I. I. (1992) Simplex-optimized and flow injection spectrophotometric assay of tetracycline antibiotics in drug formulations. Analyst, 117, 1179–1183.

Sultan S.M. and Suliman F.O. (1992) Flow-Injection spectrophotometric determination of the antibiotic ciprofloxacin in drug formulations. Analyst. 117: 1523-1526

Tamm L. K.. and Kalb E. (1993) Microspectrofluorometry on supported planar membranes. In S. G. Schulman (Ed.), Molecular luminescence spectroscopy. Methods and applications: Part 3 (pp. 303). New York, NY, USA: Wiley.

Tarafder P. K. and Thakur R. (2005) Surfactant-mediated extraction of iron and its spectrophotometric determination in rocks, minerals, soils, stream sediments and water samples. Microchemical Journal, 80, 39–43.

Teshima N., Ayukawa K.. and Kawashima T. (1996). Simultaneous flow injection determination of iron(III) and vanadium(V) and of iron(III) and chromium(VI) based on redox reactions. Talanta, 43, 1755–1760.

Teixeira L. S. G. and Rocha F. R. P. (2007) A green analytical procedure for sensitive and selective determination of iron in water samples by flow-injection solid phase spectrophotometry. Talanta, 71, 1507–1511.

Tesfaldet Z. O., van Standen J. F. and Stefan R. J. (2004) Sequential injection spectrophotometric determination of iron as Fe(II) in multi-vitamin preparations using 1,10-phenanthroline as complexing agent. Talanta, 64, 1189–1195.

Themelis D. G., Tzanavaras P. D., Kika F. S. and Sofoniou M. C. (2001) Flow-injection manifold for the simultaneous spectrophotometric determination of Fe(II) and Fe(III) using 2,2'-dipyridyl-2-pyridylhydrazone and a single-line double injection approach. Fresenius' Journal of Analytical Chemistry, 371, 364–368.

Udnan Y., Jakmunee J., Jayasavati S., Christian G. S., Synovec R. E. and Grudpan K.. (2004). Cost-effective flow injection spectrophotometric assay of iron content in pharmaceutical preparations using salicylate reagent. Talanta, 64, 1237–1240.

van Staden J. F. and Kluever L. G. (1998) Determination of total iron in ground waters and multivitamin tablets using a solid-phase reactor with tiron immobilised on amberlite ion-exchange resin in a flow injection system. Fresenius' Journal of Analytical Chemistry, 362, 319–323.

Weeks D. A. and Bruland K. W. (2002) Improved method for shipboard determination of iron in seawater by flow injection analysis. Analytica Chimica Acta, 453, 21–32.

Wirat R. (2008) Reverse flow injection spectrophotometric determination of iron(III) using chlortetracycline reagent. Talanta, 74, 1236–1241.

Yamamura S. S. and Sikes J. H. (1966) Use of citrate–EDTA masking for selective determination of iron with 1,10-phenanthroline. Analytical Chemistry, 38, 793–795.

Yan G. F., Shi G. R. and Liu Y. M. (1992) Fluorimetric determination of iron with 5-(4-methylphenylazo)-8- aminoquinoline in the presence of surfactants. Analytica Chimica Acta, 264, 121–124.

Yegorov D. Y., Kozlov A. V., Azizova O. A. and Vladimirov Y. A. (1993) Simultaneous determination of Fe(III) and Fe(II) in water solutions and tissue homogenates using desferal and 1,10-phenanthroline. Free Radical Biology & Medicine, 15, 565–574.

Yilmaz V. T. and Yilmaz F. (1999) Acetylsalicylhydroxamic acid and its cobalt(II), nickel(II), copper(II) and zinc(II) complexes. Transition Metal Chem., 24, 726-729.

8

Quality Control of Herbal Medicines with Spectrophotometry and Chemometric Techniques – Application to *Baccharis* L. Species Belonging to Sect – Caulopterae DC. (Asteraceae)

María Victoria Rodriguez[1,2], María Laura Martínez[1], Adriana Cortadi[1],
María Noel Campagna[1], Osvaldo Di Sapio[1], Marcos Derita[2],
Susana Zacchino[2] and Martha Gattuso[1]
[1]*Farmacobotánica, Área Biología Vegetal, Departamento Cs. Biológicas*
[2]*Farmacognosia, Área Farmacognosia, Departamento de Química Orgánica*
Facultad de Ciencias Bioquímicas y Farmacéuticas
Universidad Nacional de Rosario
Argentina

1. Introduction

Medicinal plants constitute a rich cultural and biological heritage in many countries, which could be very useful in meeting the therapeutic needs of the population (Rodriguez, 2010a). Traditional herbal medicines have been widely used for many years in many eastern countries (Liang et al., 2004). However, little work has been done to validate and standardize these products properly in order to match phytotherapy to chemotherapy which currently receives almost unconditional support from formal systems of health care. For several years now activities have been undertaken to systematize the identification, validation, production and use of medicinal plants, for both primary health care as well as a semi-industrial or industrial process, which implies their transformation into safe, reliable and stable phytopharmaceutical products. Therefore it is suggested that medicinal plants and their derived products would be a viable option for national development as an agricultural and therapeutic alternative, but standardization and industrialization, involving sustained yields, a quality control system and honest and reliable marketing would be needed for widespread implementation and official support. Consequently, on account of the above, education and research should be in agreement if any advance is to be made in this area (Rodriguez, 2010a). However, the necessary criteria for data quality, safety and efficacy of traditional medicine that would support its use in the world do not exist. Appropriate, accepted research methodology for evaluating traditional medicine is also lacking (Liang et al., 2004).

Most countries have developed organisms for controlling the quality of herbal remedies destined for the internal market or for export. The officially recognized drugs are subject to testing and specifications of identity, purity and contents of the principal active ingredients or markers for each plant drug, in order to guarantee the conservation of the species and the specifications of microbiological purity. Currently the quality requisites of the different national pharmacopoeias vary between countries, however the International Federation of Pharmaceutical Manufacturer Associations, have taken a first step towards the global implementation of organized criteria (International Federation of Pharmaceutical Manufacturer Associations [IFPMA], 1997). The World Health Organization has conducted a review of the legal status of traditional medicine and complementary or "alternative" therapies in 123 countries (World Health Organization [WHO], 2001). Moreover, some countries and organizations have developed, or are developing, their own monographs, for example, the Commission E of the German Ministry of Health, WHO and European Scientific Cooperative on Phytotherapy, China, Brazil and Argentina, among others. These monographs recognise the quality standards applicable to drugs and herbal remedies in the pharmaceutical market. (Blumenthal, 1998; European Scientific Cooperative on Phytotherapy [ESCOP], 2003; Keller, 1991; WHO, 1991, 1999, 2002).

In Argentina there is great interest in controlling the Quality, Security and Efficacy of herbal medicines. In order to achieve this ambitious objective the following steps must be taken: a- Registration of the products, b- Verification of good practices of production and quality control, of the crude drug and its products, and c- Pharmacovigilance. ANMAT (National Administration of Drug, Food and Medical Technology) is the organism responsible for authorizing all activities related to medicines through its technical organism, the INAME (National Institute of Drug). The products based on herbal medicine and aromatics are considered as Phytotherapeutic Drugs in Resolution N° 144/98 and the Supplementary Provisions (**Provision 2671/99; Provision 2672/99; Provision 2673/99**) that are now in effect. Provision 1788/2000 contains a list of 109 herbal medicines that can not be authorised for the production of herbal medicine on account of the contradictions or toxic effects reported in their traditional use.

It is very important that plant material is of the highest quality as it is used as a medicine. The aforementioned decrees aim to guarantee the quality of the raw material, the intermediate and finished products and a series of tests have been established among which, as a minimum requisite, are botanical identity, purity (physical-organoleptic and microbiological - for health) and their activity or composition (methodological analysis), taking into account the Good Practices of Agricultural Production and the Good Practices of Manufacturing (Rodriguez, 2010a).

1.1 Spectrophotometry and chemical fingerprints of herbal medicines

In general one or two markers or pharmacologically active compounds of herbal components or herbal blends were used to evaluate the quality and authenticity of herbal medicines in the identification of a single herb or preparation of herbal medicine and to evaluate the quantitative composition of a herbal product. However, this type of determination does not give a complete picture of the herbal products because multiple ingredients are usually responsible for its therapeutic effects. These therapeutic effects may work synergistically and change their activity on being separated into their active parts. So various analytical techniques can be applied for this type of registration and the complete herbal product can be considered as the "active compound" (Liang et al., 2004).

The concept of phytoequivalence was developed in Germany to ensure the consistency of herbal products (Tyler, 1999). According to this concept, a chemical profile, such as a spectrophotometric fingerprint for a herbal product should be constructed and compared with the profile of a clinically tested reference product. Therefore, a spectrophotometric fingerprint of a herbal medicine is a spectrophotometric pattern of an extract of some chemical components that are pharmalogically active and/or with chemical characteristics. It is suggested that with the help of spectrophotometric fingerprints obtained, the authentication and identification of herbal medicines would be appropriate, even when the amount and/or concentration of some typical chemical ingredients are not exactly the same for different samples of these herbal medicines. In this way, we are broadly considering multiple ingredients in the extracts of herbal medicines rather than one or two marker ingredients for evaluating the quality of the herbal products.

In general various analytical techniques can be used to obtain fingerprints of herbal medicines of which chromatography is most often used on account of its great efficiency in separating the different components of an extract. For the above mentioned reasons these chromatographic techniques have a high cost, especially if they are to be used for routine quality control. A simple alternative technique such as UV/Visible spectrophotometry, coupled to chemometric methods was carried out by Lonni et al. (2005) for taxonomic purposes. Authors were able to separate populations of three species of *Baccharis* according to their UV / Visible spectra. Here, we propose to apply the same methodology as part of quality control of herbal medicines.

Figure 1 shows the spectrophotometric profiles of ethanol extracts of the aerial parts of *Baccharis gaudichaudiana* DC. obtained from five areas in Misiones province, Argentina (MI1, MI2, MI3, MI4, MI5) (Rodriguez, 2010b). Here, of course, some differences can be observed in the profiles; however phytoequivalence for *B. gaudichaudiana* can be seen.

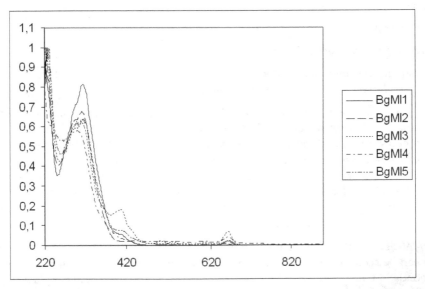

Fig. 1. Absorption UV/Visible spectra of *Baccharis gaudichaudiana* populations (Bg). MI, Misiones. Numbers indicate when there is more than one population of the same species.

1.2 Chemometric methods and quality control

The chemometric methods consist of a number of statistical, mathematical and graphic techniques that analyze many variables simultaneously (Lonni, 2005). The method used in this study is as follows:

Principal components analysis (PCA).

This method is based on the transformation of a group of original quatitative variables into another group of unrelated independent variables, known as principal components. The components have to be interpreted independently of one another, as they contain part of the variance that is not expressed in any other principal component (Pla, 1986; López & Hidalgo, 1994).

The criterion of Cliff in 1987 was adopted for selecting the number of components that should be used for the analysis, which states that the eigenvalues of acceptable components should explain 70% of the total variance (López & Hidalgo, 1994).

Proportion of variance explained: in general statistical programs provide information of the eigenvectors and, in some cases, the correlation between the original variables and the principal components. However, these correlations can be calculated from the eigenvectors in the following formula (Pla, 1986):

$$r(jk) = l(jk) \times (\lambda(\kappa))\ 1/2/s(ij) \tag{1}$$

where,

$r(jk)$ = correlation between the original variable $x(j)$ and the k-esim component.

$l(jk)$ = j-esim element of the k-esim eigenvector.

$\lambda(\kappa)$ = k-esim eigenvalue.

$s(ij)$ = variances of the correlation matrix.

In most studies it is important to determine the degree of discrimination of the variables so that those with the most and least variation can be identified.

Using PCA it is possible to determine the degree of discrimination, quantifying the proportion of variance explained by each original variable of the selected components; to do this it is necessary to add the squares of the correlation formed by each original variable with the selected components. This is possible as the components are not correlated (Pla, 1986). In the case of a variable in series: $rx12 + rx22 + rx32$ = proportion of the variance explained, having selected three components. It should to be taken into account that the variables that explain a larger proportion of variance are the most discrimatory and therefore they are more important.

The UV/Visible spectra of populations of two species, *Baccharis articulata* (Lam.) Pers. and *B. trimera* can be seen in Figure 2. In this figure it is difficult to distinguish which spectrum corresponds to which species, but the resolution is greater when PCA is applied (Figure 3) and different coordinates can be seen for the corresponding populations of one species or another.

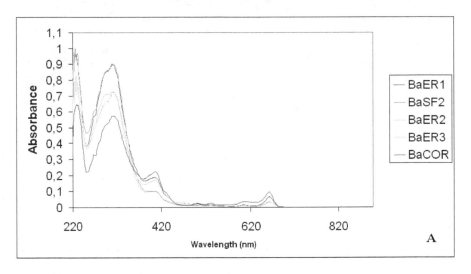

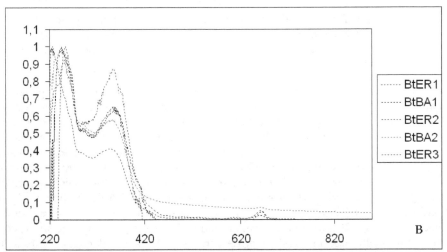

Fig. 2. Absorption UV/Visible spectra. **A**, *Baccharis articulata* (Ba); **B**, *Baccharis trimera* (Bt) populations. BA, Buenos Aires; COR, Corrientes; ER, Entre Ríos; SF, Santa Fe. Numbers indicate when there is more than one population of the same species

2. An example of differentiation of *Baccharis* L. species belonging to sect – Caulopterae DC. (Asteraceae) using UV/Visible spectrophotometry data and multivariate analysis

Baccharis L. is an exclusive American genus comprising approximately 500 species. Its distribution area covers the whole of South America and continues northwards up to the south of USA including the Atlantic coast up to Massachusetts, although its presence is greater in the intertropical and subtropical zones (Cuatrecasas, 1967; Müller, 2006). Several

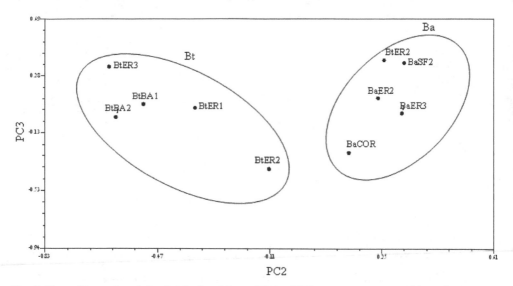

Fig. 3. Two-dimensional model derived from PCA of 700 quantitative variables of 10 *Baccharis* populations. Ba, *B. articulata*; Bt, *B. trimera*. BA, Buenos Aires; COR, Corrientes; ER, Entre Rios; SF, Santa Fe populations. Numbers indicate when there is more than one population of the same species.

authors have contributed to the infrageneric classification of *Baccharis* in general and regional floras (Ariza Espinar, 1973; Baker, 1882-1884; Barroso, 1976; Cuatrecasas, 1967; De Candolle, 1836; Giuliano, 2001; Heering, 1904; Lessing, 1831; Weddell, 1855-1856) and it was De Candolle (1836) who was the first to subdivide the genus in eight sections, mainly based on leaf morphology. More recently, Giuliano (2001) grouped 96 Argentine *Baccharis* species into 15 sections, among which the sect. Caulopterae DC. is characterized by the presence of species with alate stems. Two of nine Argentine species from this sect., i.e., *Baccharis articulata* (Lam.) Pers. and *Baccharis crispa* Spreng. are included in the National Argentine Pharmacopeia Ed. VI (1978), and a third one, *Baccharis trimera* (Less.) DC. in the Brazilian Pharmacopeia Ed. IV (2002) with the common name of "Carquejas". The nine species of this sect. are traditionally used in infusions or decoctions, as hepatic, colagogue, diuretic, ulcer healing and external antiseptics. They are also used in herbal remedies and phytotherapy and in the preparation of spirits and soft drinks (Correa, 1985; Gupta, 1995; Hieronymus, 1882; Martínez Crovetto, 1981; Sorarú & Bandoni, 1978; Toursarkissian, 1980). Beneficial effects of these species can be attributed in part to their high content of flavonoids. The chemistry of the flavonoids is predictive of their free radical scavenging activity, which confers the antioxidant activity (Harborne & Williams, 1992; Rice-Evans et al., 1995).

Currently, the morphoanatomical studies of these species in sect. Caulopterae only provide incomplete information which makes it difficult to differentiate each one properly in the non-flowering condition when the size of capitulum of each one varies (Giuliano, 2000; Müller, 2006). This fact has lead to the misuse of the same common name for botanically diverse species, which surely have different chemical compositions and therefore different pharmacological properties (Abdel-Malek et al., 1996; Desmarchelier et al., 1997; De Oliveira et al., 2003; Gene et al., 1992, 1996; Lapa et al., 1992; Palacios et al., 1983; Stoicke & Leng-Peschlow, 1987). Also the high frequency of errors committed during the collection of these species for medicinal purposes is understandable due to the coexistence of these entities in certain habitats. Hence, considering that numerical methods have been shown to be useful for multi-component metabolic classification studies, we decided to carry out population studies including 53 samples of these nine sect. Caulopterae species, combining their spectrophotometric profiles with multivariate analysis. These nine species belonging to sect. Caulopterae are: *Baccharis articulata* (Lam.) Pers., *B. crispa* Spreng., *B. gaudichaudiana* DC., *B. microcephala* (Less.) DC., *B. phyteumoides* (Less.) DC., *B. penningtonii* Heering., *B. sagittalis* (Less.) DC., *B. triangularis* Hauman and *Baccharis trimera* (Less) DC. Of these nine species, only three are official and its use is permitted but all are indiscriminately collected for medicinal purposes due to their similar morphotypes as they have alate stem.

2.1 Material and methods

Fifty three samples of nine *Baccharis* species were collected from wild materials in different locations in Argentina. All samples were botanically identified by our group and voucher specimens were deposited at the herbarium of the National University of Rosario, Argentina (Table 1).

The aerial parts of the dried plants (5 g) were macerated (24 h, 3x) with absolute ethyl alcohol. The ethanolic extract was filtered and concentrated in a rotary evaporator at a temperature lower than 100 °C. Thirty mg of dry extract were mix in 3 ml dichloromethane (DCM) and left for 1 h and then filtered with common filter paper. A dilution of 50 µl of this solution in 950 µl of methanol was prepared and filtered twice with a 0.45 µm Millipore filter (Lonni et al., 2003, 2005).

Spectrophotometric analyses were carried out using a Biochrom Model Libra S12 UV/Visible Spectrophotometer, equipped with tungsten halogen and deuterium arc light sources with a single solid state silicon photodiode detector, and operating software.

TLC analyses were carried out using silica gel 60 F254, Merck; mobile phase, DCM: Hexane:

MeOH (4:2:1). Chromatograms were evaluated under UV light at 254 and 365 nm to detect the presence of flavonoids. TLC was additionally sprayed with a diphenylborinic acid ethanolamine/polyethylene glycol reagent. Apigenin, chlorogenic acid, genkwanin, luteolin, quercetin and rutin were used as markers (purchased from Extrasynthèse, France).

HPLC analyses were carried out using a Spectra Physics Model SP8800 ternary pump chromatograph with Spectra 100 UV/Visible detector, having as chromatographic conditions, methanol eluent, 1 ml min[-1] flow, Luna C18 phenomenex (250 x 4.6 mm, 5 µm particle size). The injection volume was 100 µl and elution was monitored al 254 nm. Apigenin, genkwanin and luteolin were used as markers (purchased from Extrasynthèse, France).

TLC and HPLC analysis were applied in order to complement the studies carried out by PCA of the spectrophotometric data and to find potential markers of the species that could not be characterized by the previous method.

Sample	Voucher	Date	Sample	Voucher	Date
BaCO1	1570	Apr-05	BmFO	1656	Feb-06
BaCO2	1607	Feb-06	BmMI	1657	Feb-06
BaCO3	1617	Mar-06	BphySF1	1888	Feb-07
BaCO4	1618	Mar-06	BphySF2	1939	Mar-08
BaCO5	1619	Mar-06	BpER	1594	Jan-06
BaCO6	1620	Mar-06	BpSF1	1887	Feb-07
BaCO7	1621	Mar-06	BpSF2	1938	Mar-08
BaCOR	1903	Mar-07	BsRN1	1906	Mar-07
BaER1	1927	Apr-07	BsRN2	1953	Nov-08
BaER2	1928	Aug-07	BtrBA[a]	SI Burkart[b]	Nov-1972
BaER3	1929	Apr-07	BtrCHU[a]	SI Dacnik[b]	Feb-1969
BaSF1	1930	Aug-07	BtrLP[a]	3262	Dec-1975
BaSF2	1916	Jan-07	BtrSL	1907	Mar-07
BaSF3	1917	Jan-07	BtBA1	1955	Aug-05
BcCO1	1543	Mar-05	BtBA2	1956	Mar-05
BcCO2	1623	Mar-06	BtBA3	1668	May-06
BcCO3	1624	Mar-06	BtCOR1	1539	Mar-05
BcME1	1590	Jan-06	BtCOR2	1553	Mar-05
BcME2	1591	Jan-06	BtCOR3	1574	Jul-05
BcME3	1909	Jan-07	BtCOR4	1954	Feb-06
BcME4	1910	Jan-07	BtER1	1535	Mar-05
BgMI1	1564	Oct-05	BtER2	1537	Mar-05
BgMI2	1566	Oct-05	BtER 3	1542	Dec-04
BgMI3	1655	Feb-06	BtER4	1583	Feb-06
BgMI4	1569	Mar-05	BtER5	1645	Mar-06
BgMI5	1654	Feb-06	BtER6	1926	Apr-07
BmCOR	1572	Oct-05			

Table 1. Collection data of analysed samples of *Baccharis* species. The abbreviation mean: species: Ba, *B. articulata*; Bc, *B. crispa*; Bg, *B. gaudichaudiana*; Bm, *B. microcephala*; Bp, *B. penningtonii*; Bphy, *B. phyteumoides*; Bs, *B. sagittalis*; Btr, *B. triangularis*; Bt, *B. trimera*; provinces: BA, Buenos Aires; CHU, Chubut; CO, Córdoba; COR, Corrientes; ER, Entre Ríos; FO, Formosa; LP, La Pampa; MI, Misiones; RN, Río Negro; SF, Santa Fe; SL, San Luis. Numbers indicate when there is more than one population of the same species. [a] Material extracted from Herbarium; [b] Names of collectors, numbers not provided by the Herbarium.

2.1.1 Statistical analysis

Principal components analysis (PCA) was applied to the population study. The analysis was performed using the NTSYS-pc 2.11w (Numerical Taxonomy and Multivariate Analysis System) designed by Rohlf (1998).

Basic data matrix was prepared considering 700 absorbance values (quantitative variable) of each analysed sample (53 extracts in total).

Before PCA the data were pre-processed with normalization to unit area technique (Beebe et al., 1998).

2.2 Results and discussion

2.2.1 Spectrophotometric analysis

Seven hundred absorbance values were utilised as quantitative variables for population analysis. Samples were collected in different provinces and seasons, mainly taking into account the quantitative variability of secondary metabolites during the year (Table 1). The principal component analysis showed that the first nine components explain almost 98.81% of the total variability. The second (PC2) and the third (PC3) principal components gathered relevant information for classifying species. Figure 4, shows a two dimension plot of PC2 vs. PC3 using all the variables. Samples could be classified in five large groups containing the species *B. crispa*, *B. microcephala*, *B. phyteumoides*, *B. triangularis* and *B. trimera*. PC2 clearly separates *B. microcephala* and *B. trimera* populations from *B. crispa* and *B. phyteumoides* populations. Moreover, samples corresponding to *B. triangularis* species were separated from those belonging to *B. microcephala*, *B. trimera*, *B. crispa* and *B. phyteumoides* populations by PC2. While PC3 separates *B. microcephala* samples from the rest of the species, it also separates *B. crispa* samples from *B. phyteumoides*, *B. trimera* and *B. triangularis* and between samples of the two latter species. However, PC3 did not completely distinguish *B. phyteumoides* from *B. trimera* and *B. triangularis* samples.

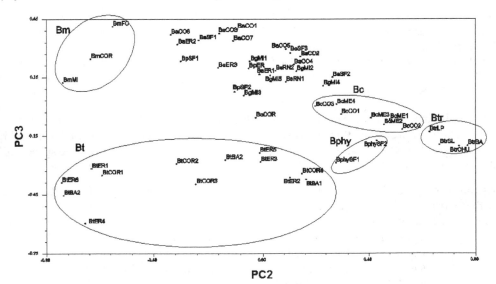

Fig. 4. Two dimensional model of PC2 *vs.* PC3 (15.63 and 7.80 %, respectively) derived from PCA of 700 quantitative variables of 53 *Baccharis* populations. Ba, *B. articulata*; Bc, *B. crispa*; Bg, *B. gaudichaudiana*; Bm, *B. microcephala*; Bp, *B. penningtonii*; Bphy, *B. phyteumoides*; Bs, *B. sagittalis*; Btr, *B. triangularis*; Bt, *B. trimera*; BA, Buenos Aires; CHU, Chubut; CO, Córdoba; COR, Corrientes; ER, Entre Ríos; FO, Formosa; LP, La Pampa; MI, Misiones; RN, Río Negro; SF, Santa Fe; SL, San Luis.

Figure 5, shows a two dimensional plot of PC7 vs. PC9 using all the variables. Here, the seventh (PC7) and ninth (PC9) principal components enable the *B. penningtonii* samples to be separated from the rest of the species.

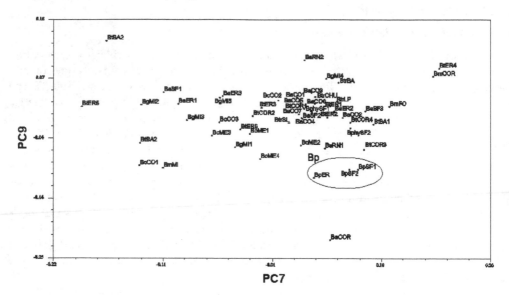

Fig. 5. Two dimensional model of PC7 *vs.* PC9 (0.62 and 0.36 %, respectively) derived from PCA of 700 quantitative variables of 53 *Baccharis* populations. Ba, *B. articulata*; Bc, *B. crispa*; Bg, *B. gaudichaudiana*; Bm, *B. microcephala*; Bp, *B. penningtonii*; Bphy, *B. phyteumoides*; Bs, *B. sagittalis*; Btr, *B. triangularis*; Bt, *B. trimera*; BA, Buenos Aires; CHU, Chubut; CO, Córdoba; COR, Corrientes; ER, Entre Ríos; FO, Formosa; LP, La Pampa; MI, Misiones; RN, Río Negro; SF, Santa Fe; SL, San Luis.

Spectrophotometric PCA data allowed the distinction of six out of the nine species examined in this study. However the three species *B. articulata*, *B. gaudichaudiana* and *B. sagittalis* could not be separated which is shown in the figure 6, where it is seen that the average spectra of the species are very similar.

As can be observed in Figure 4, the *B. microcephala* and *B. trimera* samples have PC2 scores with opposite signs to those of the *B. crispa*, *B. phyteumoides* and *B. triangularis* samples. This contrast can be explained with the help of Figure 7 A; C. Figure 7 A shows a graph of the loading values (eigenvalues) on PC2 vs. λ. Positive values are situated in a region between 200 and 230 nm. It is possible to verify from Figure 7 C that the analytical signals for *B. crispa*, *B. phyteumoides* and *B. triangularis* in this region are more intense than those of *B. microcephala* and *B. trimera*, and that the *B. crispa*, *B. phyteumoides* and *B. triangularis* samples have positive scores.

Figure 7 A also shows that three regions present negative values between: 225 and 275, 325 and 375 and 375 and 450 nm. In Figure 7 C, one can verify that in these intervals the most intense analytical signals belong to samples of the *B. microcephala* and *B. trimera* species. For this reason the *B. microcephala* and *B. trimera* species samples have negative PC2 scores.

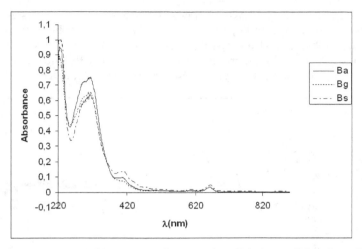

Fig. 6. Representative UV-Visible spectra of *B. articulata* (Ba), *B. gaudichaudiana* (Bg), *B. sagittalis* (Bs).

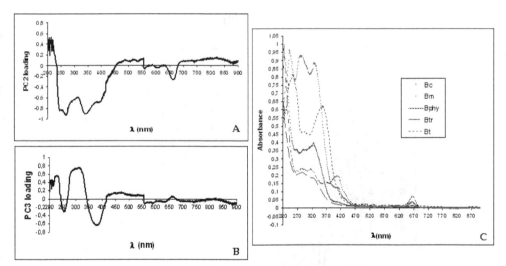

Fig. 7. A-B, PC loadings graph. **A**, PC2 (explains 15.63 % of the total data variance); **B**, PC3 (explains 7.80 % of the total data variance). C, Representative UV-Visible spectra of *B. crispa* (Bc), *B. microcephala* (Bm), *B. phyteumoides* (Bphy), *B. triangularis* (Btr), *B. trimera* (Bt).

Figure 7 B (PC3 eigenvalues) shows two regions, one with positive values (270 to 350 nm) and one with negative values (350-420 nm). From Figure 7 C it is possible to verify that in the first region the analytical signals for *B. microcephala* are more intense than those of *B. trimera*, *B. crispa*, *B. phyteumoides* and *B. triangularis*, and that the *B. microcephala* samples have positive PC3 scores. The rest of the species have negative PC3 scores. In Figure 7 B, the region with negative values (270 to 350 nm) matches the more intense analytical signals for *B. crispa*, *B. phyteumoides* and *B. triangularis* in Figure 7 C.

B. penningtonii species samples have negative PC9 scores; PC9 separates *B. penningtonii* samples from the rest as seen in Figure 5. Figure 8 A shows a graph of the loading values (eigenvalues) on PC9 vs. λ; it also shows that two regions present negative values between: 200 and 250 and 320 and 370 nm and it can be observed that the most intense analytical signals for *B. penningtonii* species samples are in these regions (Figure 8 B).

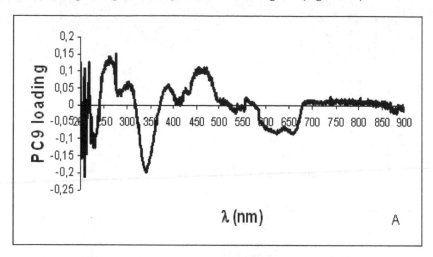

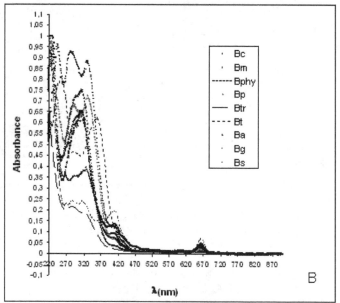

Fig. 8. A, PC loadings graph. PC9 (explains 0.36 % of the total data variance);
B, Representative UV-Visible spectra of *B. articulata* (Ba), *B. crispa* (Bc), *B. gaudichaudiana* (Bg), *B. microcephala* (Bm), *B. phyteumoides* (Bphy), *B. penningtonii* (Bp), *B. sagittalis* (Bs), *B. triangularis* (Btr), *B. trimera* (Bt).

These results suggest that the substances responsible for the discrimination between species are those which have peaks around 225, 250, 300, 325, 350, 375 and 425 nm. Considering that the ethanol extracts used in this study might possess mainly flavonoids and that the region studied is where they present their main absorption bands (Greenham et al., 2003), it is likely that the components responsible for the spectra in this work would be flavonoids. It is worth taking into account that previous studies on the phytochemistry of *Baccharis* spp. have shown the presence of flavonoids, mainly flavanones and flavones (Coelho et al., 2004; Gene et al., 1996; Torres et al., 2000). From species of this genus 298 flavonoids were isolated. Among them 24 were flavanones and 85 were flavones and among them 48% are oxygenated in the C-3 (Gonzaga Verdi et al., 2005). They were described as good chemotaxonomic markers for the lower hierarchical levels of the Asteraceae family (Emerciano et al., 2001). In contrast to other characterization studies in which it is necessary to isolate and identify chemical substances, this study uses the complete ultraviolet-visible spectra measured between 200 nm to 900 nm, avoiding the isolation, purification and characterization of chemical compounds (Lonni et al., 2005). This methodology does not distinguish the spectra of individual components in the extract. Instead, the full spectrum ranges are analyzed as a whole. Absorption spectra of figure 7 C show that the qualitative composition of flavonoids is different in the different species.

It has previously been observed that there is variability in the contents of several compounds in the different *Baccharis* species (Dresch et al., 2006; Gonzaga Verdi et al., 2005). With regards to the seasons, Borella et al. (2001) observed variations in the content of the total flavonoids, the largest number being found in a drug in the summer, which is an expected result due to the large number of functions attributed to them (Harborne & Williams, 2000). Our results are consistent with previous ones, showing different heights of peaks in the spectra in Figure 6, 7C and 8 B, even though these spectra were standardized. In a fertilization trial of *B. trimera*, in which the nutrient content of the soil was varied, no variation was observed in the contents of flavonoids (Borella et al., 2001). In our study we observed some changes in the content of flavonoids between collecting regions (Figure 4 and 5) but these did not prevent the grouping of various populations of the same species.

A routine step in multivariate data analysis is ordinarily to obtain a low-dimensional representation of the data. If two or three main components gives an accurate representation, a bi-or three-dimensional graph could be realized which mere observation is instructive. Clusters are usually easy to detect. After analysis of the eigenvalues (PC loading), more discriminant original variables are obtained. Then an ANOVA must be performing of each of these original variables between the OTUS (here species). The higher the eigenvalues, regardless of the sign will be more efficient in discriminating the OTUS. Variables that have negative eigenvalues (-) means that they are characterizing in the opposite direction in relation to the variables that have positive eigenvalues (+) and vice versa.

Thus, absorbance values of the wavelengths (225, 250, 300, 325, 350, 375 and 425 nm) obtained from the analysis of the eigenvalues were submitted to ANOVA and we have established the wavelengths that differ between pairs of species (Table 2). Thus we confirm that there are differences in the UV / Visible spectra not observable to the naked eye but are expressed as different clusters after a principal component analysis is applied.

	Ba	Bc	Bg	Bm	Bp	Bphy	Bs	Btr	Bt
Ba		300 325		250 325 350 375	250 350	300 325 350 425		225, 250, 300, 325 350	225, 250, 300, 325, 350, 375
Bc			300 325	250, 300, 325, 350 375	250 300 325 350	325 425	300 425	225 300 325 350	250 350 375
Bg				250 300 325 350	350	300 325 425		225, 250 300, 325 350	250 300 350 375
Bm					250 300	250 300 325 350	250 300 325 350	225, 250 300, 325 350, 375	300 325 375
Bp						250 300 325 350	250 350	225 250 300 325 350	250 300 325 375
Bphy							250 300 325	225 425	250, 300 325, 350 375
Bs								225 300 325 425	250 300 350 375
Btr									225, 250 325, 350 375, 425
Bt									

Table 2. Wavelengths (nm) with statistically significant differences (p <0.05) among the species. *B. articulata* (Ba), *B. crispa* (Bc), *B. gaudichaudiana* (Bg), *B. microcephala* (Bm), *B. penningtonii* (Bp), *B. phyteumoides* (Bphy), *B. sagittalis* (Bs), *B. triangularis* (Btr) and *B. trimera* (Bt)

2.2.2 TLC analysis

A TLC analysis was initiated of the dichloromethane extracts of *B. articulata*, *B. gaudichaudiana* and *B. sagittalis* used to perform UV-Visible spectra using the mobile phase DCM: Hexane: MeOH (4:2:1) in order to complement the studies carried out by PCA of the spectrophotometric data and to find potential markers of the species that could not be characterized by the previous method. NP-PEG reagent under UV 365 nm and UV 254 nm were used to detect the polyphenolic compounds present in the extracts (Wagner & Bladt, 1996), in accordance with that published for the *Baccharis* genus on account of the high occurrence of these compounds in the genus (Bohm & Stuessy, 2001; Gonzaga Verdi et al., 2005). Different color bands will be detected with the NP-PEG reagent under UV 365 nm or quenching bands will be detected under UV 254 nm if these compounds are present in the extract. On the other hand, given that one of our objectives was the identification of the species in the state of a crude drug, it was very important to select appropriate components of easy access for chemical quality control. So apigenin, genkwanin and luteolin, which are compounds present in several *Baccharis* species, were selected as markers. Genkwanin has been reported in *B. articulata* (Gianello & Giordano, 1984) and apigenin in *B. gaudichaudiana* (Fullas et al., 1994). Luteolin has not yet been reported in any of the three species analyzed by TLC, but it has been found in the following species from the same section: *B. microcephala* (Bohlmann et al., 1985), *B. trimera* (Soicke & Leng-Peschlow, 1987) and *B. triangularis* (Pettenati et al., 2007). Our results are shown in Table 3. The TLC chromatograms showed differences for the three species studied (Figures 9A and 9B). The flavonoids apigenin and genkwanin were found in all three species, although the band that corresponds to genkwanin somewhat overlaps in *B. gaudichaudiana* and *B. sagittalis*. In the case of the other marker, we observed that the band corresponding to luteolin appeared in *B. gaudichaudiana* and *B. sagittalis*. There are at least two more bands at Rf 0,83 and 0,75 for *B. articulata* and there is another band for *B. gaudichaudiana* at Rf 0,58 and three more bands for *B. sagittalis* at Rf 0,33, 0,5 and 0,58 (Figure 9 A).

	Apigenin	Genkwanin	Luteolin	Band 0,83	Band 0,75	Band 0,58	Band 0,50	Band 0,33
B. articulata	X	X	-	X	X	-	-	-
B. gaudichaudiana	X	X	X	-	-	X	-	-
B. sagittalis	X	X	X	-	-	X	X	X

Table 3. Summary of the bands obtained by TLC for *B. articulata*, *B. gaudichaudiana* and *B. sagittalis*. Mobile phase: DCM: Hexane: MeOH (4:2:1). x indicates presence of the band, - indicates absence of the band.

2.2.3 HPLC analysis

HPLC analysis was carried out on the same extracts as used for the studies with UV-Visible spectrophotometry and for TLC in *B. articulata*, *B. gaudichaudiana* and *B. sagittalis*. The chromatophotographic profiles showed the main peaks with the following retention times for *B. articulata*: 2.32, 3.00, 3.22 and 3.30 min; for *B. gaudichaudiana*: 2.26, 3.00, 3.12 and 3.30 min and for *B. sagittalis*: 2.26, 2.40, 3.00, 3.12 and 3.30 min. The retention time for the apigenin marker was 3.00 min and for genkwanin was 3.30 min. These peaks appear in all three species studied. In *B. gaudichaudiana* and *B. sagittalis* there is also a peak at 3.12 min, and this retention time corresponds to the luteolin marker (Figure 10).

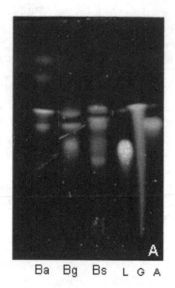

Ba Bg Bs L G A

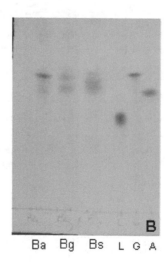

Ba Bg Bs L G A

Fig. 9. TLC of *B. articulata (Ba)*, *B. gaudichaudiana (Bg)* and *B. sagittalis (Bs)*. Chromatograms were sprayed with NP-PEG and observed under UV 365 (A) or under UV 254 without chemical treatment (B). A, apigenin; G genkwanin; L: luteolin. Mobile phase: DCM: Hexane:MeOH (4:2:1)

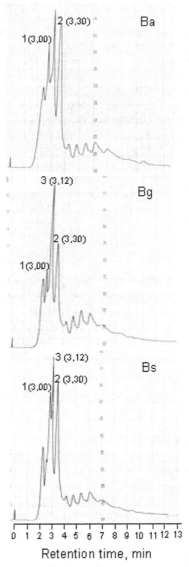

Fig. 10. Original chromatograms obtained by HPLC of *B. articulata* (Ba), *B. gaudichaudiana*
(Bg) and *B. sagittalis* (Bs),. Compounds: 1, apigenin; 2, genkwanin; 3, luteolin.

3. Conclusion

The growing tendency to use high quality, standardized plant extracts, with a guarantee of
security and efficacy, is going to continue. Therefore, all efforts should be directed at
researching the chemical-pharmalogical profiles of the extracts and combinations and in
rationalizing their therapeutic applications.

In the present example, sect. Caulopterae species are very similar between them and only *B. articulata*, *B. crispa* and *B. trimera* are official in pharmacopeia. UV/Visible spectrophotometry coupled to PCA grouped populations of these three species in different areas in a two dimensional graph. Three additional species were grouped in different areas in the same graph too (*B. microcephala*, *B. phyteumoides* and *B. penningtonii*). Populations of species that fall outside the areas of official species are unfit for medicinal purposes. It should be noted that in the case of the official species *B. articulata*, additional techniques, such as TLC should be applied to distinguish between *B. articulata*, *B. gaudichaudiana* and *B. sagittalis*. According to this, the combination of these techniques could to be used for routine quality control of official herbal medicines.

4. Acknowledgment

The present work was supported by a grant from the Agencia Nacional de Promoción Científica y Tecnológica de Argentina, PICT 1494. Dr. M V Rodriguez and Dr. Marcos Derita are research workers of CONICET.

5. References

Abdel-Malek, S., Bastien, J.W., Mahler, W.F., Jia, Q., Reinecke, M.G., Robinson, W.E., Shu, Y.H. & Zalles-Asin, J. (1996). Drug leads from the kallawaya herbalists of Bolivia. 1. Background, rationale, protocol and anti-HIV activity. *Journal of Ethnopharmacology*, Vol 50, No. 4 (March 1996), pp. 157-166, ISSN 0378-8741.

Akaike, S., Sumino, M., Sekine, T., Seo, S., Kimura, N. & Ikegami, F. (2003). A New *ent*-Clerodane Diterpene from the Aerial Parts of *Baccharis gaudichaudiana*. *Chemical & Pharmaceutical Bulletin*, Vol 51, No. 2 (February 2003), pp.197-199, ISSN 0009-2363.

Ariza Espinar, L. (1973). Las especies de *Baccharis* (Compositae) de Argentina Central. *Boletín de la Academia Nacional de Ciencias*, Vol 50 (Septiembre 1973), pp.175-305, ISSN 0325-2051.

Baker, J.G. (1882-1884). Compositae III. Asteroideae, Inuloideae. IV. Heliathoideae, Anthemideae, Senecionideae, Cynaroideae, Ligulatae, Mutisiaceae, In: *Flora brasiliensis: enumeratio plantarum 6 (3)*, C.F. Martius, A.W. Eichler, (Eds.), 1-100, F. Fleischer, Leipzig, Munich.

Barroso, G.M. (1976). Compositae, Subtribo Baccharidinae Hoffman. Estudo das espécies ocorrentes no Brasil. *Rodriguésia*, Vol 28, pp.3-273.

Borella, J.C., Fontoura, A., Menezes, Jr.A. & França, S.C. (2001). Influência da adubação mineral (N-P-K) e sazonalidade no rendimento e teor de flavonóides em indivíduos masculinos de *Baccharis trimera* Less. (Asteraceae) – Carqueja. *Revista Brasileira de Plantas Medicinais*,Vol 4, No.1 (Outubro 2001), 101-104, ISSN 1516-0572.

Bohlmann, F., Banerjee, S., Jalupovic, J., Grenz, M., Misra, L.N., Hirschmann, G.S., King, R.M. & Robinson, H. (1985). Clerodane and labdane diterpenoids from *Baccharis* species. *Phytochemistry*, Vol 24, No. 3 (October 1985), pp. 511-515, ISSN 0031-9422.

Bohm, B.A. & Stuessy, T.F. (2001) *Flavonoids of the Sunflower Family (Asteraceae)*. Springer, Wien, New York.

Blumenthal, M. (1998). *The complete german commission E monographs. Herbal guide to herbal medicines*, American Botanical council, Austin.

Coelho, M.G.P., Reis, P.A., Gava, V.B., Marques, P.R., Gayer, C.R., Laranja, G.A.T., Felzenswalb, I. & Sabino, K.C.C. (2004). Antiarthritic effect and subacute toxicological evaluation of *Baccharis genistelloides* aqueous extract. *Toxicology Letters*, Vol 154, No. 1-2 (December 2004), pp. 69-80, ISSN 0378-4274.

Correa, M.P. (1985). *Dicionário das plantas úteis do Brasil e das exóticas cultivadas*, IBDF, Río de Janeiro, Brasil.

Cuatrecasas, J. (1967). Revisión de las especies colombianas del género *Baccharis. Revista de la Academia Colombiana de Ciencias Exactas*, Vol 13, pp. 5-102.

De Candolle, A.P. (1836). Compositae: *Baccharis. Prodromus Systematis Naturalis Regni Vegetabilis*, Vol 5, pp. 398-429.

De Oliveira, S.Q., Dal-Pizzol, F., Gosmann, G., Guillaume, D. Moreira, J.C. & Schenckel, F.P. (2003). Antioxidant activity of *Baccharis articulata* extracts: isolation of a new compound with antioxidant activity. *Free Radical Research*, Vol 37, No. 5 (January 2003), pp.555-559, ISSN 1071-5762.

Desmarchelier, G., Bermúdez, M.J.N., Coussio, J., Ciccia, G. & Boveris, A. (1997). Antioxidant and prooxidant activities in aqueous extracts of argentine plants. *International Journal of Pharmacognosy*, Vol 35, No.2 (January 1997), 116-120, ISSN 1388-0209.

Dresch, A.P., Montanha, J.A., Matzenba, N.E. & Mentz, C.A. (2006). Controle De Qualidade De Espécies Do Gênero *Baccharis* L.(Asteraceae) Por CCD A Partir De Extratos Rápidos. *Infarma*, Vol 18, No. 11-12, pp. 37-40, ISSN 0104-0219.

Emerciano, V.P., Militão, J.S.L., Campos, C.C., Romoff, P., Kaplan, M.A.C., Zambon, M. & Brant, A.J.C. (2001). Flavonoids as chemotaxonomic markers for Asteraceae. Biochem. *Biochemical Systematic and Ecology*, Vol 29, No. 9 (October 2001), pp.947-957.

ESCOP (European Scientific Cooperative on Phytotherapy) (2003). *ESCOP Monographs. The scientific foundation for herbal medicinal products*, ESCOP, Exeter; Georg Thieme Verlag, Stuttgart; Thieme NewYork, New York.

Farmacopea Nacional Argentina VI ed. (1978). *Codex Medecamentarius Argentino*. Buenos Aires, Argentina.

Farmacopéia Brasileira IV Edição. (2002). Oficializada Governo Federal. Atheneu Editora S. A., Sao Paulo LTDA.

Fullas, F., Hussain, R.A., Chai, H., Pezzuto, J.M., Soejarto, D.D. & Kinghorn, A.D. (1994). Cytotoxic constituents of *Baccharis gaudichaudiana. Journal of Natural Products*, Vol 57, No. 6 (June 1996), pp.801-807, ISSN 0163-3864.

Gené, R.M., Marín, E. & Adzet, T. (1992). Anti-Inflammatory Effect of Aqueous Extracts of Three Species of the Genus *Baccharis. Planta Medica*, Vol 58 (January 1992), pp.565-566, ISSN 0032-0943.

Gené, R.M., Cartañá, C., Adzet, T., Marín, E., Parella, T. & Cañigueral, S. (1996). Anti-Inflammatory and Analgesic Activity of *Baccharis trimera*: Identification of its Active Constituents. *Planta Medica*, Vol 62, pp. 232-235, ISSN 0032-0943.

Gianello, J.C. & Giordano, O.S. (1984). Exámen químico en seis especies del género *Baccharis*. *Revista Latinamericana de Quimica*, Vol 15, pp.84-86, ISSN 0370-5943.

Giuliano, D.A. (2001). Clasificación Infragenérica de las especies argentinas de *Baccharis* (Asteraceae, Astereae). *Darwiniana*, Vol 39, No. 1-2, pp. 138-154, ISSN 0370-5943.

Gonzaga Verdi, L., Costa Brighente, I.M. & Pizzolatti, M.G. (2005). Gênero Baccharis (Asteraceae): Aspectos Químicos, Econômicos e Biológicos. *Química Nova*, Vol 28, No.1, pp. 85-94, ISSN 0100-4042.

Gupta, M.P. (1995). *270 Plantas medicinales iberoamericanas*. Programa Iberoamericano de Ciencia y Tecnología para el Desarrollo, CYTED. Convenio Andrés Bello Santafé de Bogotá, Colombia.

Greenham, J., Harborne, J.B. & Williams, C.A. (2003). Identification of Lipophilic Flavones and Flavonols by Comparative HPLC, TLC and UV Spectral Analysis. *Phytochemical Analysis*, Vol 14, No. 2 (March/April 2003) pp. 100-118, ISSN 1099-1565.

Harborne, J.B. & Williams, C.A. (2000). Advances in flavonoid research since 1992. *Phytochemistry*, Vol 55, No. 6 (November 2000), pp. 481-504, ISSN 0031-9422.

Heering, W.C. (1904). Die *Baccharis*-Arten des Hamburger Herbars. *Jahrbuch der Hamburgischen Wissenschaftlichen Anstalten*, Vol 21, pp. 1-46.

Hieronymus, J. (1882). Plantae Diaforicae, Florae Argentinae. Tomo VI. *Boletín de la Academia Nacional de Ciencias*, Vol 4, pp. 159-160, ISSN 0325-2051.

IFPMA (International Federation of Pharmaceutical Manufacturers Associations) (1997). Major steps towards global drugs regulations. Brussels.

Keller, K. (1991). Legal requirements for the use of phytopharmaceutical drugs in the Federal Republic of Germany. *Journal of Ethnopharmacology*, Vol 32, pp. 225-229, ISSN 0378-8741.

Lapa, A.J., Fischman, L.A. & Gamberini, M.T. (1992). Inhibitors of gastric secretion from Brazilian folk medicinal plants, In: *Natural Drugs and the Digestive Tract*, F. Capasso, N. Mascolo (Eds.), 63-68, EMSI, Roma.

Lessing, C.F. (1831). Synanthereae: Molins-Alatae. *Linnaea*, Vol 6, pp. 83-170.

Liang, Y., Peishan, X. & Chan, K. (2004) Quality control of herbal medicines. *Journal of Chromatography B*, Vol 812 (September 2004), pp. 53-70, ISSN 1570-0232.

Lonni, A.A.S.G., Scarminio, I.S., Silva, L.M.C. & Ferreira, D.T. (2003). Differentiation of Species of the Baccharis Genus by HPLC and Chemometric Methods. *Analytical Sciences*, Vol 19, pp. 1013-1017, ISSN 0910-6340.

Lonni, A.A.S.G., Scarminio, I.S., Silva, L.M.C. & Ferreira, D.T. (2005). Numerical Taxonomy Characterization of *Baccharis* Genus Species by Ultraviolet-Visible Spectrophotometry. *Analytical Sciences*, Vol 21, pp. 235-239, ISSN 0910-6340.

López, J.A. & Hidalgo, M.D. (1994) Análisis de Componentes Principales y Análisis Factorial, En: *Fundamentos de Estadística con Systat*, M. Ato, J.J. López, (Eds.), 457-503, Addison Wesley Iberoamericana.

Martínez Crovetto, R. (1981). *Plantas utilizadas en medicina en el NO de Corrientes*. Fund. Miguel Lillo, Miscelán, 69, Tucumán.

Müller, J. (2006). *Systematics of Baccharis (Compositae, Astereae) in Bolivia, including an overview of the genus.* (Systematics Botany Monographs v. 76). The American Society of Plant Taxonomists, ISBN 0912861762, Michigan.

Palacios, P., Gutkind, G., Randina, R.V.D., De Torres, R. & Coussio, J.D. (1983). Antimicrobial activity of *B.crispa* and *B. notosergila*. Genus Baccharis II. *Planta Medica*, Vol 49, pp. 128, ISSN 0032-0943.

Petenatti, E.M., Petenatti, M.E., Cifuente, D.A., Gianello, J.C., Giordano, O.S., Tonn, C.E. & Del Vitto, L.A. (2007). Medicamentos Herbarios en el Centro-Oeste Argentino. VI. Caracterización y Control de Calidad de dos Especies de "Carquejas": *Baccharis sagittalis* y *B. triangularis* (Asteraceae). *Latin American Journal of Pharmacy*, Vol 26 No. 2, pp. 201-208, ISSN 0326-2383.

Pla, L.E. (1986) *Análisis Multivariado: Método de Componentes Principales.* Secretaría de la Organización de Estados Americanos (OEA), Washington, D.C.

Rice- Evans, C.A., Miller, N.J. & Paganga, G. (1996). Structure-Antioxidant Activity Relationships of Flavonoids and Fenolic Acids. *Free Radical Biology and Medicine*, Vol 20, pp. 933-956, ISSN 0891-5849.

Rodriguez M.V. (2010a). Caracterización y Normalización de un recurso vegetal autóctono medicinal: especies del género *Baccharis* L. Facultad de Ciencias Bioquímicas y Farmacéuticas, Universidad Nacional de Rosario, Argentina.

Rodriguez M.V., Martínez M. L., Cortadi A. A., Bandoni A., Giuliano D. A., Gattuso S. J. & Gattuso M. A. (2010b). Characterization of three sect. Caulopterae species (*Baccharis*-Asteraceae) inferred from morphoanatomy, polypeptide profiles and spectrophotometry data. *Plant Systematic and Evolution*, Vol 286, No. 3-4 (June 2010), pp. 175-190, ISSN 0378-2697.

Rohlf, F.J. (1998). On applications of geometric morphometrics to studies of ontogeny and phylogeny. *Systematic Biology*, Vol 47, pp. 147-158, ISSN 1063-5157.

Sorarú, S.B. & Bandoni, A.L. (1978). *Plantas de la Medicina Popular Argentina.* Albatros, Buenos Aires.

Stoicke, H. & Leng-Peschlow, E. (1987). Characterization of flavonoids from *Baccharis trimera* and their antihepatotoxic properties. *Planta Medica*, Vol 53, pp. 37-39, ISSN 0032-0943.

Torres, L.M.B., Gamberini, M.T., Roque, N.F., Lima-Landman, M.T., Souccar, C. & Lapa, A.J. (2000). Diterpene from *Baccharis trimera* with a Relaxant Effect on Rat Vascular Smooth Muscle. *Phytochemistry*, Vol 55, No. 6 (November 2000), pp. 617-619, ISSN 0031-9422.

Toursarkissian, M. (1980). *Plantas Medicinales de la Argentina.* Hemisferio Sur S. A., Buenos Aires.

Tyler, V.E. (1999). Phytomedicines: Back to the future. *Journal of Natural Products*, Vol 62, pp. 1589-1592, ISSN 0163-3864.

Wagner, H. & Bladt, S. (1996) *Plant Drug Analysis. A Thin Chromatography Atlas.* Springer-Verlag, Berlin.

Weddell, H.A. (1855-1856). *Chloris andina*, vol. 1. P. Bertrand, Paris.

WHO (World Health Organization) (1991). Guidelines for the Assessment of herbal medicines. WHO/TRM/91.4, Ginebra.

WHO (World Health Organization) (1999). WHO monographs on selected medicinal plants. Vol 1, Ginebra.

WHO (World Health Organization) (2001). Legal status of traditional medicine and complementary alternative medicine: A world wide review. WHO/EDM/TRM/2001.2, Geneva.

WHO (World Health Organization) (2002). WHO monographs on selected medicinal plants. Vol 2, Ginebra.

Permissions

The contributors of this book come from diverse backgrounds, making this book a truly international effort. This book will bring forth new frontiers with its revolutionizing research information and detailed analysis of the nascent developments around the world.

We would like to thank Jamal Uddin, for lending his expertise to make the book truly unique. He has played a crucial role in the development of this book. Without his invaluable contribution this book wouldn't have been possible. He has made vital efforts to compile up to date information on the varied aspects of this subject to make this book a valuable addition to the collection of many professionals and students.

This book was conceptualized with the vision of imparting up-to-date information and advanced data in this field. To ensure the same, a matchless editorial board was set up. Every individual on the board went through rigorous rounds of assessment to prove their worth. After which they invested a large part of their time researching and compiling the most relevant data for our readers. Conferences and sessions were held from time to time between the editorial board and the contributing authors to present the data in the most comprehensible form. The editorial team has worked tirelessly to provide valuable and valid information to help people across the globe.

Every chapter published in this book has been scrutinized by our experts. Their significance has been extensively debated. The topics covered herein carry significant findings which will fuel the growth of the discipline. They may even be implemented as practical applications or may be referred to as a beginning point for another development. Chapters in this book were first published by InTech; hereby published with permission under the Creative Commons Attribution License or equivalent.

The editorial board has been involved in producing this book since its inception. They have spent rigorous hours researching and exploring the diverse topics which have resulted in the successful publishing of this book. They have passed on their knowledge of decades through this book. To expedite this challenging task, the publisher supported the team at every step. A small team of assistant editors was also appointed to further simplify the editing procedure and attain best results for the readers.

Our editorial team has been hand-picked from every corner of the world. Their multi-ethnicity adds dynamic inputs to the discussions which result in innovative outcomes. These outcomes are then further discussed with the researchers and contributors who give their valuable feedback and opinion regarding the same. The feedback is then collaborated with the researches and they are edited in a comprehensive manner to aid the understanding of the subject.

Apart from the editorial board, the designing team has also invested a significant amount of their time in understanding the subject and creating the most relevant covers. They scrutinized every image to scout for the most suitable representation of the subject and create an appropriate cover for the book.

The publishing team has been involved in this book since its early stages. They were actively engaged in every process, be it collecting the data, connecting with the contributors or procuring relevant information. The team has been an ardent support to the editorial, designing and production team. Their endless efforts to recruit the best for this project, has resulted in the accomplishment of this book. They are a veteran in the field of academics and their pool of knowledge is as vast as their experience in printing. Their expertise and guidance has proved useful at every step. Their uncompromising quality standards have made this book an exceptional effort. Their encouragement from time to time has been an inspiration for everyone.

The publisher and the editorial board hope that this book will prove to be a valuable piece of knowledge for researchers, students, practitioners and scholars across the globe.

List of Contributors

Vagner de Alencar Arnaut de Toledo and Emerson Dechechi Chambó
Animal Science Department, Brazil

Maria Claudia Colla Ruvolo-Takasusuki
Cell Biology and Genetics Department, Brazil

Arildo José Braz de Oliveira and Sheila Mara Sanches Lopes
Pharmacy Department, Universidade Estadual de Maringá, Maringá, Paraná Universidade
Estadual de Maringá, Maringá, Brazil

Ioan Țăranu, Bogdan-Ovidiu Țăranu and Iuliana Popa
National Institute of Research-Development for Electrochemistry and Condensed Matter,
Romania

Zoltan Szabadai, Vicențiu Vlaia and Lavinia Vlaia
University of Medicine and Pharmacy "Victor Babeş", Romania

Joanna Karpinska
Institute of Chemistry, University of Bialystok, Bialystok, Poland

Vesna Weingerl
University of Maribor, Faculty of Agriculture and Life Sciences, Hoče, Slovenia

Anna Błażewicz
Department of Analytical Chemistry, Medical University of Lublin, Poland

Vladimir V. Shelkovnikov
Novosibirsk Institute of Organic Chemistry SB RAS, Novosibirsk, Russia

Alexander I. Plekhanov
Institute of Automation and Electrometry SB RAS, Novosibirsk, Russia

Ibrahim Isildak
Yildiz Technical University, Faculty of Chemical and Metallurgical Engineering,
Bioengineering Department, İstanbul, Turkey

**María Laura Martínez, Adriana Cortadi, María Noel Campagna, Osvaldo Di Sapio and
Martha Gattuso**
Farmacobotánica, Área Biología Vegetal, Departamento Cs. Biológicas, Argentina

Marcos Derita and Susana Zacchino
Farmacognosia, Área Farmacognosia, Departamento de Química Orgánica, Facultad de Ciencias Bioquímicas y Farmacéuticas, Universidad Nacional de Rosario, Argentina

María Victoria Rodriguez
Farmacobotánica, Área Biología Vegetal, Departamento Cs. Biológicas, Argentina
Farmacognosia, Área Farmacognosia, Departamento de Química Orgánica, Facultad de Ciencias Bioquímicas y Farmacéuticas, Universidad Nacional de Rosario, Argentina

Printed in the USA
CPSIA information can be obtained
at www.ICGtesting.com
JSHW011404221024
72173JS00003B/420

9 781632 384249